Environmental Health Engineering in the Tropics

Second Edition

Environmental Health Engineering in the Tropics

An Introductory Text
Second Edition

Sandy Cairncross
UNICEF/WHO Interagency Team for
Guinea Worm Eradication
Ouagadougou, Burkina Faso

and

Richard G. Feachem
Dean, London School of Hygiene and
Tropical Medicine, London, England

JOHN WILEY & SONS
Chichester - New York - Brisbane - Toronto - Singapore

Other Wiley Editorial Offices

John Wiley & Sons, Inc., 605 Third Avenue,
New York, NY 10158-0012, USA

Jacaranda Wiley Ltd, G.P.O. Box 859, Brisbane,
Queensland 4001, Australia

John Wiley & Sons (Canada) Ltd, 22 Worcester Road,
Rexdale, Ontario M9W 1L1, Canada

John Wiley & Sons (SEA) Pte Ltd, 37 Jalan Pemimpin #05-04,
Block B, Union Industrial Building, Singapore 2057

Library of Congress Cataloging-in-Publication Data
Cairncross, Sandy.
    Environmental health engineering in the tropics : an introductory
text / Sandy Cairncross and Richard G. Feachem. — 2nd ed.
        p.   cm.
    Includes bibliographical references and index.
    ISBN 0 471 93885 8
    1. Sanitary engineering — Tropical conditions.   2. Sanitary
engineering — Developing countries.   I. Feachem, Richard G., 1947–
    II. Title.
TD126.5.C35   1993
628′.0913 — dc20
                                                    92-41230
                                                    CIP

British Library Cataloguing in Publication Data

A catalogue record for this book is available from the British Library

ISBN 0 471 93885 8

Typeset by Laser Words, Madras
Printed in Great Britain by Redwood Books, Trowbridge, Wiltshire

Contents

Part IV Environmental Modifications and Vector-borne Diseases

Preface

ENVIRONMENTAL HEALTH ENGINEERING AND POVERTY

The word 'tropics' in the title of this book refers only very loosely to the hotter regions of the world. Most hot countries are poor countries, and most poor countries have hot climates. The particular problems of engineering and of disease which are discussed in these pages are not confined to the tropics; many of them apply to all poor communities, whether tropical or temperate. For example, it has become clear in recent years that environmental health in many rural areas of Eastern Europe is no better than can be found in parts of Latin America, or even Africa.

Environmental health engineering is an important weapon in the fight against disease. Many of the dramatic improvements in public health in the developed countries over the last century or so are attributable to public health engineers, at least as much as to doctors. But these improvements have yet to reach the vast majority of the world's poor, who still endure the high rates of death, disease, and disability which have always been associated with poverty.

It has become increasingly apparent in recent years that these effects of poverty may be alleviated without a substantial increase in a country's per capita income. There are now a number of countries, such as the People's Republic of China and Sri Lanka, where, although the per capita income remains low, such indices of the quality of human life as the infant mortality rate, life expectancy, and rates of infection with certain parasites, have improved significantly.

Our book is concerned with engineering methods for improving the health of the poorest sections of the world's population. It is not the rich who suffer from the diseases we discuss, nor is it the rich who are deprived of the environmental services and facilities documented here. The rich in a developing country do not suffer much from schistosomiasis, their children do not frequently die of diarrhoea and malnutrition, they generally drink clean water and have hygienic excreta disposal facilities. It is the poor who suffer from these diseases and who live in conditions of environmental squalor.

The need for this book, then, and indeed the need for the consideration of tropical environmental health engineering as a distinct subject, derives directly from the needs of the poor. There is no point in offering technology to the poor which is appropriate for the rich and industrialized countries, because the poor will be unable to afford it. Nor is there any point in waiting for that distant day when per capita incomes are sufficiently high that the poor can afford to buy the services which the citizens of North America and Europe have long enjoyed. Instead one must try to make available to the poor improvements in their environment and in their life style which they can afford now, and thereby seek to improve their quality of life despite their low incomes.

An improvement in the quality of life of the poorest sections of the world's population rests essentially on two requirements. The first, and most important, is political will. Without a political commitment to distribute resources equitably among the population and provide for the basic needs of the poor, there can be little prospect of change for the poorest sections of any society. If we look at the countries in which, although poor, the majority of the population has a relatively good quality of life, we generally find them to be countries where this political will is strong and where it has been strong for a number of years.

The second requirement is the one to which this book is addressed. That is the availability of the know-how and the technology to bring improvements and facilities to the poor at a cost which they and their governments can afford. In some fields the basic scientific and engineering understanding exists and the technology is well tested, although often not widely used in practice. In other fields, such as environmental methods for diarrhoeal disease control, our understanding is less good and there is a need for fundamental scientific research on which new technological advances can be based. It goes without saying that this research must be poverty-oriented, meaning that it must be directed at problems which are experienced by the poor and must seek solutions which the poor can afford.

Engineers whose work involves environmental health in a poor country rapidly find that they must concern themselves with a variety of issues for which they are not traditionally trained. On the medical side, the diseases of the poor in hot climates are very different from the diseases of the rich in temperate climates, and public health engineers must understand these diseases if they are to do their job properly. The economic picture, as has been said above, is also radically different and public health engineers in a developing country must have a fundamental grasp of the economic climate in which they are working. Social and institutional problems are of the utmost importance, and it is necessary that projects are designed in the full knowledge of the institutional and social context in which they will be carried out. A broad range of skills and abilities is therefore required and a high degree of interdisciplinary collaboration is necessary.

This collaboration is likely to require the participation of engineers, doctors, biologists, economists, anthropologists, and others. It is our hope that this book will provide a useful introduction for those — whatever their specialist discipline — with the will and interest to confront the problems of environmental health in poor communities.

Sandy Cairncross
Richard Feachem

PREFACE TO THE SECOND EDITION

In bringing this book up to date, I have tried to incorporate the main lessons learned from the International Drinking Water and Sanitation Decade. Most of these are not technical developments; though we have learned something about subjects such as composting, waste re-use and small bore sewers, these do not involve radically new techniques. Indeed the main technical advance of the Decade has been increased recognition for the low-cost solutions which the first edition sought to promote.

The most important lessons have concerned policy and strategies for implementing water and sanitation programmes, and I have tried to summarise these in Chapters 4 and 7. We have also learned to give more attention to aspects of environmental health other than water supply and excreta disposal, and this understanding lies behind the addition of a new chapter on drainage. Above all, the bold assertions we made in the first edition are now supported by much more published documentation; I have therefore overhauled the reference list at the and of each chapter to help the reader find it.

I would like to thank Ursula Blumenthal and Peter Kolsky for helpful comments, and Mimi Khan and Caroline Smart for secretarial help. Finally, thanks to Joy Barrett for drawing my attention to the fact that (thankfully!) public health engineers are not all male.

Sandy Cairncross
Ouagadougou, 1993

Acknowledgements

This book contains a distillation of much knowledge and experience and many people have contributed to it in various ways. We wish to thank our colleagues and students, past and present, many of whom may find their own ideas and approaches in these pages. We are especially grateful for the advice of David Bradley, Nicholas Greenacre, Duncan Mara, Warren Pescod, John Pickford, Geoffrey Read, David Simpson, Brian Southgate, and Graham White. Dr Cairncross' work in preparing the second edition was supported by the UK Overseas Development Administration. Finally, we are grateful to Lois and Zuzana for their encouragement and support.

Part I
Health and Pollution

1

Engineering and Infectious Disease

1.1 INTRODUCTION

Environmental health engineering concerns engineering methods for the improvement of the health of the community. In practice it has come to focus especially upon domestic water supplies and excreta disposal facilities. However, other engineering interventions, such as drainage, housing, and irrigation scheme design, are also relevant in the tropics.

One is unlikely to succeed in improving community health unless one understands it, and for a public health worker in the tropics this understanding must chiefly apply to the infectious diseases. Although most of these have now been brought under control or even eliminated in the industrialized countries, they are still the principal cause of ill-health among the world's poor.

An infectious disease is one which can be transmitted from one person to another or, sometimes, to or from an animal. All infectious diseases are caused by living organisms, such as bacteria, viruses, or parasitic worms, and a disease is transmitted by the passing of these organisms from one person's body to another.

During the transmission process, the organisms may be exposed to the environment, and their safe passage to the body of a new victim is then vulnerable to changes in the environment. Environmental health engineering therefore seeks to modify the human environment in such a way to prevent or reduce the transmission of infectious diseases.

In considering the transmission of infectious disease, the distinction between the state of being infected, and the state of being ill, must be borne in mind. Very often, the section of the population most involved in transmitting an infection shows little or no sign of disease; conversely, people who are seriously ill with the disease may be of little or no importance in transmission. For example, when a cholera epidemic sweeps through a community, those who show signs of cholera are only a small minority of those who are infected with the disease and likely to pass it on to others.

For the environmental health engineer, it is convenient to start by classifying the relevant infectious diseases into categories which relate to the various aspects of the environment which he can alter. The conventional generic or biological taxonomy[1] of disease classifies them according to the nature of their pathogens[2] — the organisms which cause them. This classification system is unhelpful to environmental interventions. For instance, smallpox is grouped with hepatitis and filariasis[3] is grouped with Guinea worm. In this chapter, infectious diseases are grouped in a way which is more helpful to the environmental health worker.

1.2 WATER-RELATED INFECTIONS

A water-related disease is one which is in some way related to water or to impurities in water. It is necessary to distinguish the infectious water-related diseases from those related to some chemical property of the water. Damage to the teeth and bones, for instance, is associated in some countries with high fluoride levels; this non-infectious type of water-related disease is dealt with in Chapter 2. In developing countries it is usually the infectious water-related diseases which are of prime importance; these are the subject of this chapter.

Classification of transmission mechanisms

Before we can classify the water-related infections, we must define the four distinct types of water-related route by which a disease may be transmitted from one person to another. These are shown in Table 1.1, and are related there to the environmental strategies for disease control which are appropriate to each. The four routes are described below.

1. Water-borne route   Truly water-borne transmission occurs when the pathogen is in water which is drunk by a person or animal which may then become infected. Potentially water-borne diseases include the classical infections, notably cholera and typhoid, but also include a wide range of other diseases, such as infectious hepatitis, diarrhoeas, and dysenteries.

The term 'water-borne disease' has been and still is, greatly abused so that it has become almost synonymous with 'water-related disease'. It is essential to use the term water-borne only in the strict sense defined here.

---

[1] Appendix A is a guide to the conventions of biological species classification.
[2] Words such as this are included in the Glossary in Appendix B.
[3] Appendix C is a checklist of relevant diseases.

Table 1.1   The four types of water-related transmission route for infections, and the preventive strategies appropriate to each

Transmission route	Preventive strategies
Water-borne	Improve quality of drinking water Prevent casual use of unprotected sources
Water-washed (or water-scarce)	Increase water quantity used Improve accessibility and reliability of domestic water supply Improve hygiene
Water-based	Reduce need for contact with infected water[1] Control snail populations[1] Reduce contamination of surface waters[2]
Water-related insect vector	Improve surface water management Destroy breeding sites of insects Reduce need to visit breeding sites Use mosquito netting

[1] Applies to schistosomiasis only.
[2] The preventive strategies appropriate to the water-based worms depend on the precise life-cycle of each (see Appendix C) and this is the only general prescription that can be given.

Another source of misunderstanding has been the assumption that, because a disease is labelled water-borne this describes its usual, or even its only means of transmission. The preoccupation with strictly water-borne transmission has its origins in the dramatic water-borne epidemics of cholera and typhoid which occurred in some European towns in the last century and the first quarter of this one, and were largely caused by urban water supplies with inadequate treatment facilities. Similar epidemics sometimes occur in tropical towns today, but *all* water-borne diseases can also be transmitted by any route which permits faecal material to pass into the mouth (a 'faecal–oral' route). Thus cholera may be spread by various faecal–oral routes, for instance via contaminated food. Water-borne transmission is merely the special case of drinking faecal material in water, and any disease which can be water-borne can also be transmitted by other faecal–oral routes.

2. Water-washed route   There are many infections of the intestinal tract and of the skin which, especially in the tropics, may be significantly reduced following improvements in domestic and personal hygiene. These improvements in hygiene often hinge upon increased availability of water and the use for hygienic purposes of increased volumes of water. Their transmission can be described as 'water-washed'; it depends on the quantity of water used, rather than the quality. The relevance of water to these diseases is that it is an aid to hygiene and cleanliness, and its quality is relatively unimportant for this purpose. A water-washed disease may be formally defined as one whose transmission will be reduced following an increase in the volume of water used for hygienic purposes, irrespective of the quality of that water (Figure 1.1).

Water-washed diseases are of three main types. Firstly, there are infections of the intestinal tract, such as diarrhoeal diseases, which are important causes of serious illness and death especially among young children in poor countries. These include cholera, bacillary dysentery, and other diseases previously mentioned under water-borne diseases (Figure 1.2). These diseases are all faecal–oral in their transmission route and are therefore potentially either water-borne or water-washed. Any disease which is transmitted by the pathogen passing out in the faeces of an infected person and subsequently being ingested (a faecal–oral disease) can either be transmitted by a truly water-borne route or by an almost infinite number of other faecal–oral routes, in which case it is probably susceptible to hygiene improvements and is therefore water-washed. A number of investigations have shown that diarrhoeal diseases, especially bacillary dysentery (shigellosis), decreased with the availability of water and with the volume of water used but did not decrease significantly with improvements in the microbiological quality of the water. The conclusion is that these diarrhoeal diseases, although potentially water-borne, were in fact primarily water-washed in the communities studied, and were mainly transmitted by faecal–oral routes which did not involve water as a vehicle.

The second type of water-washed infection is that of the skin or eyes. Bacterial skin sepsis, scabies, and fungal infections of the skin are extremely prevalent in many hot climates, while eye infections such as trachoma are also common and may lead to blindness. These infections are related to poor hygiene and it is to be anticipated that they will be reduced by increasing the volume of water used for personal hygiene. However, they are quite distinct from the intestinal

Figure 1.1 Hand-washing is an important method of interrupting water-washed transmission. Hand-washing is especially important after defaecation, before eating, and before preparing food. A community health nurse in Ghana is teaching children about the importance of hand-washing (photo: WHO)

Figure 1.2 Cholera and other acute watery diarrhoeas can kill people. Young children are especially at risk. These people died of cholera in Calcutta, India. Acute diarrhoeas can often be treated by encouraging the patient to drink a solution of sugar and salt—a technique known as oral rehydration
(Photo: D Henrioud, WHO)

water-washed infections because they are not faecal–oral and cannot be water-borne. They therefore relate primarily to water quantity and are not significantly related to water quality.

The third type of water-washed infection is also not faecal–oral and therefore can never be water-borne. These are infections carried by lice which may be reduced by improving personal hygiene and therefore reducing the probability of infestation of the body and clothes with these arthropods. Louse-borne epidemic typhus (due to infection by *Rickettsia prowazeki*) is mainly transmitted by body lice, which cannot persist on people who regularly launder their clothes. (Note that these are not the same as head lice.) Louse-borne relapsing fever (due to infection by a spirochaete, *Borrelia recurrentis*) may also respond to changes in hygiene linked to increased use of water for washing (see Chapter 15).

3. Water-based route A water-based disease is one whose pathogen spends a part of its life cycle in a water snail or other aquatic animal. All these diseases are due to infection by parasitic worms (helminths) which depend on aquatic intermediate hosts to complete their life cycles. The degree of sickness depends upon the number of adult worms which are infecting the patient and so the importance of the disease must be measured in terms of the intensity of infection as well as the number of people infected. An important example is schistosomiasis in which water, polluted by excreta, contains aquatic snails in which the schistosome worms develop until they are shed into the water as infective cercariae and re-infect man through his skin (see Chapter 17).

Another water-based disease is Guinea worm (*Dracunculus medinensis*), which is found in most of West Africa (Figure 1.3), and has a unique transmission route (Appendix D). The mature female worm, about 0.5 m long, lies under the skin, usually on the leg, and creates a painful blister. When this blister is immersed in water or water is splashed onto it, as is often done to soothe the pain, the worm releases thousands of microscopic larvae. If the larvae are washed into a pond or shallow well, they are eaten by cyclopoids, which then become infected, and they develop inside these new hosts. Cyclopoids are tiny crustaceans that are found in many small bodies of water. They are only 0.8 mm long, and so are easily consumed inadvertently in water from an infected pond or well. Infected cyclopoids tend to sink to the bottom, so the risk is greatest when only a shallow depth of water remains. The cyclopoids themselves are not dangerous to drink, but any *Dracunculus* worms they contain will develop further in the human host and any fertilized female worm will make her way to the legs and form a new blister a year later, ready to start a new cycle.

Although a water-based disease, Guinea worm is the only

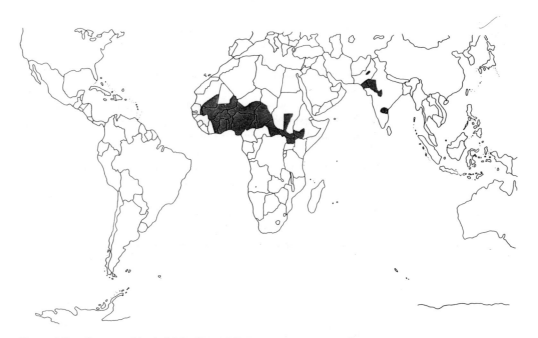

Figure 1.3   Geographical distribution of Guinea worm, *Dracunculus medinensis*. The areas marked are those in which Guinea worm was reported or probably existed in 1991.
Source: Data from Centres for Disease Control, Atlanta

infection which is exclusively transmitted in drinking water. It kills few people but causes debilitating pain, usually in the planting season, and so can have far-reaching economic effects by reducing the ability to work of most of the population. The complete eradication of Guinea worm has been adopted as a goal by the World Health Organization. Several endemic countries have already achieved considerable progress towards it, and it may well be achieved by the end of this century.

Provision of safe water supplies is the primary measure to control the disease, especially in communities where 20% or more of the population is regularly infected. When the annual incidence has been reduced to 10% or less, most of the cases are likely to be caused by the casual use of infected water sources away from the home, while working in the fields, visiting or travelling. In such cases, or in very small villages where safe water supplies are not yet affordable, the inhabitants must be encouraged to filter their drinking water through a cloth to remove the cyclopoids, to avoid drinking unsafe water, and to ensure that people with blisters do not enter drinking-water sources.

The other diseases in this category are acquired by eating insufficiently cooked fish, crabs, crayfish, or aquatic vegetation; they are clearly unrelated to water supply, but they may be affected by excreta disposal (see below).

4. Insect vector route   The fourth and final route is for water-related diseases to be spread by insects which either breed in water or bite near water. Malaria, yellow fever, dengue, and onchocerciasis (river blindness), for example, are transmitted by insects which breed in water while West African sleeping sickness is transmitted by the riverine tsetse fly (*Glossina* spp.) which bites near water. Insect-borne diseases are discussed in Chapter 15.

Classification of infections

Table 1.1 lists these four water-related transmission routes and links them to their appropriate preventive strategies. In order that these concepts may be employed to assess the impact on health of a water improvement scheme, it is necessary first to list the chief water-related diseases and assign them to an appropriate category. However, all the faecal–oral infections can be transmitted by both water-borne and water-washed routes, so they are placed in a special category of their own (Table 1.2). The second category is reserved for infections that are exclusively water-washed; in other words, the skin and eye infections plus diseases which are associated with lice. Each water-related infection can then be assigned to one of the following four categories: (1) faecal–oral: (2) water-washed; (3) water-based; and (4) insect-vectored. Table 1.2 lists the major water-related infections and assigns them to their category in addition to linking them to the type of organism which causes them.

Table 1.2   Environmental classification of water-related infections

Category	Infection	Pathogenic agent
1. Faecal-oral	Diarrhoeas and dysenteries	
(water-borne or	Amoebic dysentery	P
water-washed)	Balantidiasis	P
	Campylobacter enteritis	B
	Cholera	B
	Cryptosporidiosis	P
	E. coli diarrhoea	B
	Giardiasis	P
	Rotavirus diarrhoea	V
	Salmonellosis	B
	Shigellosis (bacillary dysentery)	B
	Yersiniosis	B
	Enteric fevers	
	Typhoid	B
	Paratyphoid	B
	Poliomyelitis	V
	Hepatitis A	V
	Leptospirosis	S
2. Water-washed:		
(a) skin and eye	Infectious skin diseases	M
infections	Infectious eye diseases	M
(b) other	Louse-borne typhus	R
	Louse-borne relapsing fever	S
3. Water-based:		
(a) penetrating skin	Schistosomiasis	H
(b) ingested	Guinea worm	H
	Clonorchiasis	H
	Diphyllobothriasis	H
	Fasciolopsiasis	H
	Paragonimiasis	H
	Others	H
4. Water-related insect vector		
(a) biting near water	Sleeping sickness	P
(b) breeding in	Filariasis	H
water	Malaria	P
	River blindness	H
	Mosquito-borne viruses	
	Yellow fever	V
	Dengue	V
	Others	V

B = Bacterium   R = Rickettsia
H = Helminth   S = Spirochaete
P = Protozoon   V = Virus
M = Miscellaneous
See Appendix C for further details.

1.3 EXCRETA-RELATED INFECTIONS

All the diseases in the faecal–oral category mentioned above, as well as most of the water-based diseases and several others not related to water, are caused by pathogens transmitted in human excreta, normally in the faeces. In a similar manner to the water-related diseases, the classification of these excreta-related diseases can help us to understand the effects on them of the various possible engineering solutions to the problem of excreta disposal.

Those of the excreta-related diseases which are also water-related can of course be controlled, at least partially, by improvements in water supply and hygiene. But these and the other excreta-related diseases may also be affected by improvements in excreta disposal, ranging from the construction or improvement of toilets to the choice of methods for transport, treatment, and final disposal or re-use of excreta. To understand the effects of excreta disposal on these diseases, a further classification is required (Table 1.3).

I. Faecal–oral diseases (non-bacterial) Improvements in excreta disposal will have differing degrees of influence on the various faecal–oral diseases. Some of these infections, caused by viruses, protozoa, and helminths, can spread very easily from person to person and are transmitted in affluent communities in Europe and North America which have high standards of sanitary facilities and hygiene. Changes in excreta disposal methods are unlikely to have much effect on their incidence unless accompanied by sweeping changes in personal cleanliness, requiring substantial improvements in water supply, as well as major efforts in health education.

II. Faecal–oral diseases (bacterial) For the faecal–oral diseases caused by bacteria, person-to-person transmission route are important but so too are other routes with longer transmission cycles, such as the contamination of food, crops, or water sources with faecal material. Some of the pathogens in this category, notably *Campylobacter, Cryptosporidium, Salmonella,* and *Yersinia* are also passed in the faeces of animals and birds (see Figure 1.4; animals in parallel). This suggests that they will not be greatly influenced by improvements in excreta disposal unless animals' excreta are also removed.

III. Soil-transmitted helminths This category contains several species of parasitic worm whose eggs are passed in faeces. They are not immediately infective, but first require a period of development in favourable conditions, usually in moist soil. They then reach their next human host by being ingested, for instance on vegetables, or by penetrating the soles of the feet. Since the eggs are not immediately infective, personal cleanliness has little effect on their transmission, but any kind of latrine which helps to avoid faecal contamination of the floor, yard, or fields will limit transmission. However, if a latrine is poorly maintained with an earth floor which becomes soiled, it

Table 1.3   Environmental classification of excreta-related infections

Category	Infection	Pathogenic agent	Dominant transmission mechanisms	Major control measures (engineering measures in italics)
I Faecal-oral (non-bacterial) Non-latent, low infectious dose	Poliomyelitis	V	Person to person contact	*Domestic water supply*
	Hepatitis A	V	Domestic contamination	*Improved housing*
	Rotavirus diarrhoea	V		*Provision of toilets*
	Amoebic dysentery	P		Health education
	Giardiasis	P		
	Balantidiasis	P		
	Enterobiasis	H		
	Hymenolepiasis	H		
II Faecal-oral (bacterial) Non-latent, medium or high infectious dose, moderately persistent and able to multiply	Diarrhoeas and dysenteries		Person to person contact	*Domestic water supply*
	Campylobacter enteritis	B	Domestic contamination	*Improved housing*
	Cholera	B	Water contamination	*Provision of toilets*
	E. coli diarrhoea	B	Crop contamination	*Excreta treatment prior to re-use or discharge*
	Salmonellosis	B		Health education
	Shigellosis	B		
	Yersiniosis	B		
	Enteric fevers			
	Typhoid	B		
	Paratyphoid	B		
III Soil-transmitted helminths Latent and persistent with no intermediate host	Ascariasis (roundworm)	H	Yard contamination	*Provision of toilets with clean floors*
	Trichuriasis (whipworm)	H	Ground contamination in communal defaecation area	*Excreta treatment prior to land application*
	Hookworm	H		
	Strongyloidiasis	H	Crop contamination	

IV Beef and pork tapeworms Latent and persistent with cow or pig intermediate host	Taeniasis	H	Yard contamination Field contamination Fodder contamination	*Provision of toilets* *Excreta treatment prior to land application* Cooking and meat inspection
V Water-based helminths Latent and persistent with aquatic intermediate host(s)	Schistosomiasis Clonorchiasis Diphyllobothriasis Fasciolopsiasis Paragonimiasis	H H H H	Water contamination	*Provision of toilets* *Excreta treatment prior to discharge* *Control of animals harbouring infection* Cooking
VI Excreta-related insect vectors	Filariasis (transmitted by *Culex pipiens* mosquitoes) Infections in Categories I–V, especially I and II, which may be transmitted by flies and cockroaches	H M	Insects breed in various faecally contaminated sites	*Identification and elimination of potential breeding sites* Use of mosquito netting

B= Bacterium   V = Virus
H= Helminth   M= Miscellaneous
P = Protozoon
See Appendix C for further details.

can become a focus for transmission. Dirty latrines may then cause more transmission than would occur if people were to defaecate in widely scattered locations in the bush.

The eggs of these worms can survive for months between hosts, so that treatment of excreta is vital if they are to be re-used on the land. The eggs can be eliminated by sedimentation in stabilization ponds (Chapter 10), by the heat of composting (Chapter 13), or by prolonged storage.

IV. Beef and pork tapeworms   These tapeworms of the genus *Taenia* require a period in the body of an animal host before re-infecting man when the animal's meat is eaten without sufficient cooking. Any system which prevents untreated excreta being eaten by pigs and cattle will control the transmission of these parasites. Adequate treatment is required where sewage or sludge is applied to grazing land (Chapter 14). These infections have to pass through an animal before returning to infect another person. This transmission by animals in series is different from transmission in parallel; in the latter, the pathogen infects man and animals in a similar way, and the faeces of both may transmit it to either man or animal (Figure 1.4).

V. Water-based helminths   All of the water-based diseases already mentioned, except for Guinea worm, are caused by helminths which are passed in excreta and must then pass a stage in the body of an aquatic host, usually a snail (Appendix C). They then re-infect man through the skin or when insufficiently cooked fish, crabs, crayfish, or aquatic vegetation are eaten. Appropriate excreta disposal methods can help to control them by preventing untreated excreta from reaching water in which the aquatic hosts live. However, in all cases except *Schistosoma mansoni* and *S. haematobium*, transmission by animals in parallel may occur (Figure 1.4). Animals' faeces are therefore a source of infection so that measures restricted to human excreta can have only a partial effect. Moreover, since one worm can multiply in the snail host to produce thousands of larvae, faecal contamination must be practically eliminated in order to reduce transmission. This is very hard to achieve.

VI. Excreta-related insect vectors   These are of two main kinds. First the *Culex pipiens* group of mosquitoes, found throughout most of the world, breeds in highly polluted water, for instance in septic tanks and flooded pit latrines, and transmits filariasis in some regions (see Chapter 15). Second, the flies and cockroaches which breed where faeces are exposed. They carry pathogenic organisms on their bodies and in their intestinal tracts. Their nuisance value is great, and flies can also spread excreted pathogens and eye infections.

Figure 1.4   Two forms of zoonosis. Some infections affect only man and are transmitted by various routes from person to person. Cholera is an example. Other infections also affect other vertebrate animals and may be transmitted from animals to man and vice versa. Two types of transmission cycle followed by zoonoses are shown in the figure. When animals are in series, the disposal of human excreta may greatly reduce transmission; when animals are in parallel, human excreta disposal will not affect transmission from animals to man.
Source: From Feachem *et al.* (1983)

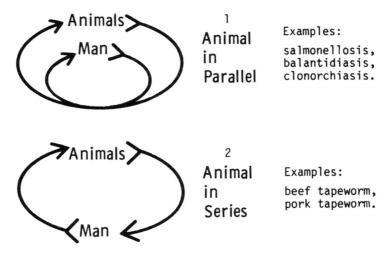

1
Animal in Parallel

Examples:
salmonellosis, balantidiasis, clonorchiasis.

2
Animal in Series

Examples:
beef tapeworm, pork tapeworm.

Latency and persistence

The pathogens of categories III–V of excreta-related diseases (Table 1.3) show a property known as latency. This means that they cannot infect man immediately after they have been excreted, but must first undergo a period of development in soil, pigs, cows, or aquatic animals. Another important characteristic of each pathogen is its persistence — how long it can survive in the environment. These factors are illustrated graphically in Figure 1.5. As might be expected, the latent and more persistent organisms have 'longer' transmission cycles, and the efficacy of improved sanitation in controlling them depends on this cycle (see Figure 1.6).

Conclusions

The potential impact of sanitation improvements, and of improvements in personal hygiene, on the various categories of excreta-related disease, is summarized in Table 1.4. For most of these diseases, an improvement in excreta disposal is only one of several measures required for their control. It is essential that people of all ages use the improved toilets and keep them clean. The disposal of children's excreta is at least as important as that of adults (Figure 1.7). Studies in the past have often failed to detect beneficial effects from improved sanitation because, although latrines were built they were not kept clean and were not used by children, or by adults when working in the fields (see Chapter 7).

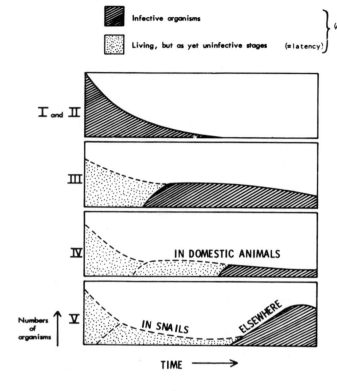

Figure 1.5 The survival of pathogens in the environment by category of excreta-related infection (Table 1.3)
Source: From Feachem *et al.* (1983)

1.4 REFUSE-RELATED INFECTIONS

Poor refuse disposal will encourage fly-breeding and may thus promote the transmission of faecal–oral infections as described above. It can also promote diseases associated with rats, such as plague, leptospirosis, salmonellosis, endemic typhus and rat-bite fever. Uncollected refuse can obstruct streets and drainage channels (Figure 1.8).

Culex quinquefasciatus mosquitoes breed in the wastewater retained in such blocked channels, and may transmit filariasis. Other

Table 1.4 Potential for control of excreta-related infections by improvements in sanitation and personal hygiene

Disease category from Table 1.3	Impact of sanitation alone	Impact of personal hygiene alone
I Non-bacterial faecal-oral	Negligible	Moderate
II Bacterial faecal-oral	Slight to moderate	Moderate
III Soil-transmitted helminths	Great	Negligible
IV Beef and pork tapeworms	Great	Negligible
V Water-based helminths	Moderate	Negligible
VI Insect vector	Slight to moderate	Negligible

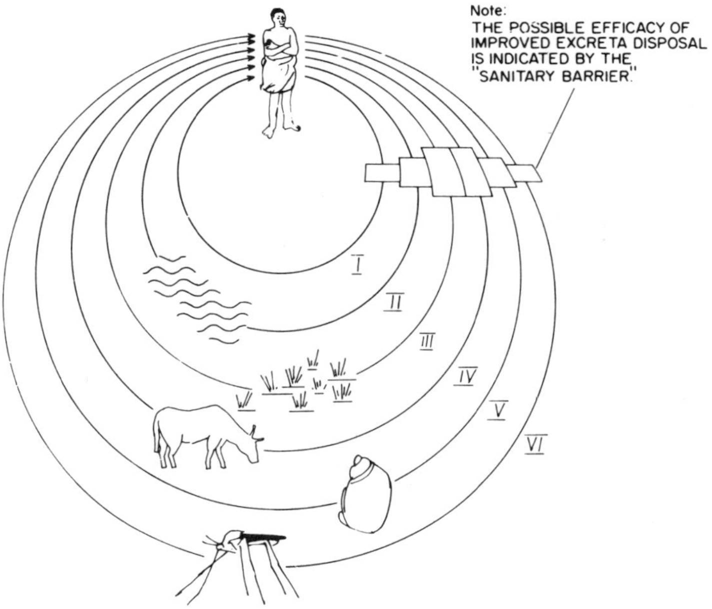

Note:
THE POSSIBLE EFFICACY OF
IMPROVED EXCRETA DISPOSAL
IS INDICATED BY THE
"SANITARY BARRIER."

Figure 1.6 The length and dispersion of the transmission cycles of six categories of excreta-related infection (Table 1.3). The possible efficacy of improved excreta disposal in interrupting transmission is also shown
Source: From Feachem *et al.* (1983)

Figure 1.7 Children are the main sufferers from most excreted infections and the faeces of young children frequently contain excreted pathogens. The hygienic disposal of the faeces of babies and young children is therefore very important to protect the health of the family and the community. After defaecation, babies and young children should be washed. (Photo: D Ayres)

species of mosquito, breeding in the refuse itself (tin cans, old tyres, etc.) can spread viral infections such as dengue and yellow fever.

1.5 HOUSING-RELATED INFECTIONS

The interactions between housing and human health are numerous; five major areas will be mentioned here (Figure 1.9).

Figure 1.8 A lane running between the backs of houses is completely blocked by refuse in this Nigerian town. Municipal vehicles are denied access for either refuse collection or emptying of latrines (Photo: D Mara)

First, the location of housing can have important effects on the health of the inhabitants. This is particularly relevant to vector-borne diseases such as malaria or sleeping sickness (see Chapter 15), where housing built close to high vector concentrations may increase disease transmission.

Second, the manner in which the house design and location promotes or hinders domestic hygiene will have bearing on all diseases related to domestic hygiene. These are all the faecal–oral infections and all the water-washed infections; in other words Categories 1 and 2 in Table 1.2 and Categories I and II in Table 1.3.

Third, housing has an influence on airborne infections: measles, mumps, meningitis, diphtheria, all respiratory infections, and pneumonic plague. Housing design will affect crowding, ventilation, air temperature, and humidity, all of which will affect the transmission of airborne pathogens. A smoke-filled or otherwise irritating atmosphere will also influence the susceptibility of individuals to respiratory infections. However, it is not normally possible to demonstrate a decisive association between larger, better ventilated rooms and better health. While it is clear that overcrowding and close physical contact present many opportunities for transmission of airborne infections, so also do many events outside the house such as crowded buses or markets.

Fourth, the manner in which the house promotes or discourages populations of rats, insects or domestic animals will influence the prevalence of all infections related to them. In Latin America, for example, poor housing encourages infestation with the bugs which are the vectors of Chagas' disease (Chapter 15). In general, any housing which people share with animals or poultry may assist the transmission of diseases carried by those animals and their parasites.

Figure 1.9   These four pictures illustrate a wide variety of housing conditions. You might consider the potential health hazards associated with each. In general, urban housing is very much worse than rural housing. (a) Rural housing in Chad (Photo: WHO). (b) Temporary rural housing for migrant cotton-pickers in the Sudan (Photo: E Schwab, WHO). (c) An urban slum in India (Photo: E Schwab, WHO). (d) Housing over water in the Philippines (Photo: WHO)

Certain details of house design can have an important influence on diseases transmitted by night-biting mosquitoes. A false ceiling, or mosquito netting across the eave space between walls and roof, can be a very effective improvement as this is where mosquitoes often enter.

Lastly, earth floors in houses can harbour the eggs and larvae of intestinal worms (Table 1.3, Category III). They are also conducive to certain parasitic insects such as the blood-sucking floor maggot (*Auchmeromyia luteola*) and the sand-flea or jigger (*Tunga penetrans*), both of which are widespread in Africa. African tick-borne relapsing fever, caused by *Borrelia duttoni*, is spread by the tick *Ornithodorus moubata* which hides during the day in the dust and cracks of earth floors and mud walls, emerging to feed at night.

1.6 SUMMARY

Appendix C gives a tabulated summary of the principal water-related and excreta-related infections. Readers may find it useful to have a short medical dictionary or reference book at hand to provide further details of infections with which they are not familiar. Benenson (1990) is particularly recommended for this purpose.

1.7 REFERENCES AND FURTHER READING

Benenson, A.S. (1990). *Control of Communicable Diseases in Man*. 15th edition (Washington: American Public Health Association).

Esrey, S.A., Feachem, R.G. and Hughes, J.M. (1985). Interventions for the control of diarrhoeal diseases among young children: improving water supplies and excreta disposal facilities. *Bulletin of the World Health Organization*, **63**, 757–772.

Feachem, R.G., Bradley, D.J., Garelick, H. and Mara, D.D. (1983). *Sanitation and Disease: Health Aspects of Excreta and Wastewater Management* (London: John Wiley).

Feachem, R.G., McGarry, M.G. and Mara, D.D. (1977). *Water, Wastes and Health in Hot Climates* (London: John Wiley).

Hardoy, J.E., Cairncross, S. and Satterthwaite, D. (1990). *The Poor Die Young; Housing and Health in Third World Cities* (London: Earthscan Publications).

Jeffrey, H.C. and Leach, R.M. (1975). *Atlas of Medical Helminthology and Protozoology* (Edinburgh: Churchill Livingstone).

Martin, A.E., Kaloyanova, F. and Maziarka, F. (1976). *Housing, the Housing Environment and Health: an Annotated Bibliography*, Offset Publications, No. 27 (Geneva: World Health Organization).

White, G.F., Bradley, D.J. and White, A.U. (1972). *Drawers of Water; Domestic Water Use in East Africa* (Chicago: University of Chicago Press).

WHO (1987). *Prevention and Control of Intestinal Parasitic Infections*, Technical Report Series No. 749 (Geneva: World Health Organization).

2

Health and Water Chemistry

2.1 INTRODUCTION

In Chapter 1, a classification of infectious diseases related to water is outlined. This chapter describes briefly the health problems related to the chemical, as opposed to the microbiological, composition of water. The sight, smell, taste, and even the feel of water is affected by chemicals contained within it. The chemistry of water can lead to disease either if there is an absence of a necessary constituent or, more commonly, if there is an excess of a harmful chemical. These diseases are clearly not infectious and are prevented simply by adding the chemical which is deficient or removing the chemical which is harmful. For communities which take water from a piped and treated source, this presents no major obstacles in that chemicals may be added or removed at the treatment works. The benefits of increasing or reducing a given compound must be weighed against the extra treatment costs involved.

Communities that use untreated supplies, such as most villages in developing countries, face more serious problems if there is a chemical problem associated with their water source. If they lack a necessary chemical it would be extremely difficult to add it to the water and, since the water is untreated, it is impossible to remove a harmful chemical pollutant. In addition, chemical water pollution may lead to an unpleasant taste or appearance and this may cause people to abandon certain sources in favour of others which are more acceptable to them. This is potentially serious if it causes the rejection of a ground-water source of good microbiological quality in favour of a surface-water source of poor microbiological quality.

Water chemistry and disease may be considered under three headings: the absence of necessary chemicals, the excess of harmful organics, and the excess of harmful inorganics.

2.2 THE ABSENCE OF NECESSARY CHEMICALS

The absence of essential substances in water is not generally a problem because there are alternative sources of these substances in food. Iodine deficiency in water, for instance, has to occur in association with iodine deficiency in the diet before widespread goitre is likely to result. Endemic goitre occurs among isolated rural communities in several areas of the world, but is best controlled by the introduction of iodized salt and injections of iodized oil.

A deficiency of fluoride in water and diet can cause poor growth of bones and teeth in the young. Communities with low overall fluoride intake may experience a higher incidence of dental caries. For this reason, the addition of up to 1 mg/l of fluoride to public water supplies at the treatment works is now common practice in several developed countries. In communities not having a treated drinking water supply, it would not usually be feasible to make up a deficiency in fluoride. Fluoride deficiency is likely to be a low priority among the health problems of the area and, if the administration of supplementary fluoride is considered desirable, it may be better introduced in the diet.

A statistical association has been found in some developed countries between water softness and heart disease. In general, populations using hard water (containing the carbonates and sulphates of calcium and magnesium) have a lower incidence of cardiovascular disease. Research continues and, as yet, there is no acceptable explanation of this phenomenon. Until an explanation is found, it is unlikely that there will be planned changes in water quality in order to try to affect cardiovascular disease incidence. Heart disease is in any case much less a problem in most developing countries than it is in Europe and North America.

Other relationships have been postulated between certain trace elements in water and resistance to disease. For example, the low chromium content in the diet and drinking water of a village in Jordan was found to cause juvenile diabetes. But such cases are rare and the evidence is often inconclusive.

In conclusion, then, the absence of necessary chemical constituents in water is not a widespread problem among the poor in hot climates. In particular areas where iodine or fluoride deficiencies occur, the remedy is as likely to be in the diet as the water. An exception to this is communities relying on desalinated sea water, as in several countries in the Middle East and a few small islands in the Caribbean and elsewhere. Desalinated sea water may have to be reconstituted with the required chemicals by the addition of natural water, or of measured chemicals, in order to make its taste acceptable to the consumer.

2.3 HARMFUL ORGANICS

Over a thousand different organic compounds have now been identified in water. Many of these (micropollutants) occur in very small concentrations, often less that 1 $\mu g/l$, and are not yet known to have any specific health effects. However, some organic compounds, or groups of compounds, are known to be either toxic, or carcinogenic (cancer-producing) or to produce odours or tastes, sometimes after reacting with chlorine used for disinfection.

Most of the toxic substances are pesticides (including herbicides, fungicides, insecticides, and molluscicides) which are applied in very large quantities in some regions of the world. The use of the more common compounds is monitored by the United Nations Food and Agriculture Organization (FAO) and the World Health Organization (WHO) and tolerable daily intakes have been determined. The bulk of pesticide application is in agriculture and thus most human intake is in food and not water. Moreover, settlement in still water or in sedimentation tanks (Chapter 6) will reduce the concentration of most pesticides. However, in areas where there is heavy runoff from treated fields into surface waters, or where water has been treated directly with insecticides or molluscicides, there is a danger that pesticide levels in surface water used for drinking may exceed permissible limits, whether or not it has been filtered or chlorinated. The technology for detecting and controlling pesticides in water is complex and it is necessary to have a central laboratory in each country or region keeping an eye on this problem.

Most surface waters in developing countries are used to catch fish. Any pesticides present in the water are likely to accumulate in the food chain, and particularly in bottom-feeding species of fish, presenting a greater hazard to the people who eat them than to anyone drinking the water. If there is cause for concern, therefore, the first step should be to test for pesticides in the fish.

Some other organic chemicals are known to cause cancer when consumed in large doses and to occur in minute concentrations in drinking water supplies. In the USA the use of granular activated carbon treatment is recommended to reduce organic compounds in public drinking water supplies. Particular attention has been paid to polynuclear aromatic hydrocarbons (PAHs) and to trihalomethanes (THMs). WHO (1984) recommend that concentration of one representative PAH in drinking water (benzo[a]pyrene) should not exceed 0.01 $\mu g/l$, although food, not water, contributes almost 99% of people's total PAH intake. As with pesticides, central and sophisticated laboratory facilities are required to detect and identify these chemicals in drinking water.

Among the many organic compounds found in drinking water, greatest attention has been paid to the trihalomethanes (THMs) — especially chloroform. THMs may be present in concentrations of 1–100 $\mu g/l$ and occasionally drinking water contains more than 100 μg/l of chloroform. THMs are mainly formed during water treatment by the reaction of chlorine with 'precursor' organics in the raw water. THMs are carcinogenic in laboratory animals at high and sustained doses, but there is little concrete evidence of risks to humans associated with low levels of exposure via drinking water. The many other risk factors in cancer need to borne in mind and it has been pointed out that a few spoonfuls of some brands of cough linctus provide a chloroform intake greater than a lifetime's consumption of most drinking waters. While the USA has proposed a total trihalomethane (THM) limit of 0.1 mg/l, other countries have not followed suit and there is little or no scientific basis for establishing a THM standard in drinking water at present.

Most countries, and certainly all developing countries, may safely take the view that the presence of THMs and most other organic pollutants in drinking waters is of negligible public health importance. It certainly cannot justify failure to disinfect drinking water. When concern over THMs led the authorities in Lima, Peru, to stop chlorinating local wells, it was followed by the 1991 cholera epidemic which claimed over 3500 lives.

2.4 HARMFUL INORGANICS

Of more importance than the organic compounds are the harmful inorganic chemicals sometimes found in water. A number of metallic ions are known to cause metabolic disturbances in man by upsetting the production and function of certain enzymes, or to cause a variety of other toxic effects. Antimony, arsenic, barium, beryllium, boron, cadmium, cobalt, lead, mercury, molybdenum, selenium, tin, uranium, and vanadium are all implicated although the acceptable intakes and physiological effects are not known for all of these. Table 2.1 sets out WHO-recommended limits for nine health-related substances.

However, it is important to bear in mind the proportion of the total intake of a particular substance which is likely to be contained in water. In the case of certain elements, the contribution of water is likely to be minor when compared with the intake in food. This principle is illustrated in Table 2.2 which gives tentative figures, for the United Kingdom, of the likely contribution of drinking water to the total human intake of various elements.

Much publicity has been given to the possibility of links between aluminium concentrations in drinking water and the incidence of Alzheimer's Disease, a form of senile dementia which is not only found in the elderly. However, a recent report by the US Food and Drug Administration concluded that, 'The evidence linking aluminium to Alzheimer's Disease is not strong ... as there have been no conclusive findings as yet.'

Table 2.1   Guideline values for health-related inorganic constituents in drinking water (*Source*: From WHO (1984))

Substance	Concentration (mg/l)
arsenic	0.05
cadmium	0.005
chromium	0.05
cyanide	0.1
fluoride	1.5[1]
lead	0.05
mercury	0.001
nitrate (as N)	10.0
selenium	0.01

[1]Guideline value may vary, depending on climate and water consumption

Table 2.2   Possible contribution of trace elements from drinking water to total human intake in the United Kingdom

< 1%	1–5%	5–10%	10–20%
arsenic	beryllium	aluminium	barium
iron	cadmium	boron	lead
molybdenum	chlorine	calcium	lithium
potassium	chromium	copper	strontium
selenium	magnesium		
silicon	manganese		
sodium	mercury		
titanium	nickel		
	silver		
	sulphur		
	tin		
	zinc		

Source: From Commins (1978). Reproduced by permission of B.T. Commins, W.R.C.

A more significant problem in developing countries is the effect of salts in ground water — mainly chlorides and sulphates — in making the water unpalatable, and so leading people to use surface water which is more likely to be bacteriologically polluted. These salts tend to make the water undrinkable before they reach seriously harmful concentrations. For instance, there are some communities in North Africa and Iran which regularly drink water containing as much as 3000 mg/l of chlorides and 1500 mg/l of sulphates; such high doses would make the water quite unpalatable and would have a laxative effect on those who were not used to them.

More recently, evidence has accumulated linking a high intake of sodium, usually as sodium chloride, with high blood pressure. But drinking water usually contributes only a minority of the total intake of salt in the diet, and the problem only arises when the water is so salty that it tastes bad. Moreover, it has been shown that people who rate the taste of their drinking water as bad consume considerably less than those assessing the taste as good (WHO Regional Office for Europe, 1979).

Although a deficiency of fluoride is implicated in tooth decay (Section 2.2), concentrations of over 2 mg/l of fluorides have been associated with mottling of tooth enamel, and concentrations over about 4 mg/l, consumed for many years, may cause stiffness and pain in the joints and skeletal deformities, particularly in hot climates where people drink more water, where concentrations tend to be increased by evaporation of stored water, and where people's diets may also be rich in fluoride or nutritionally deficient. The technical problem of fluoride removal is discussed in Chapter 6.

It also seems that fluoride can contribute towards Genu Valgum, another type of skeletal deformity chiefly associated with a high dietary intake of molybdenum, as it is only found in areas prone to fluorosis. It is a crippling knock-kneed condition found in a few communities in India since irrigation raised their ground-water level and increased the uptake of molybdenum by their sorghum crops. Fluoride and molybdenum together may exacerbate the effects of dietary deficiencies, particularly of copper (Krishnamachari and Krishnaswamy, 1974). This illustrates the complex interactions between various substances found in water and in food, and also the way in which a population's total intake of them may be affected by other engineering interventions, such as irrigation schemes.

Nitrate concentrations over 45 mg/l in drinking water are potentially hazardous to health in two ways. The nitrates are reduced in the body to nitrites and can cause a serious blood condition in infants known as methaemoglobinaemia (infantile cyanosis), particularly if their diet is not rich in vitamin C. It has also been suggested that very high nitrate concentrations, such as those found in some community water supplies in Colombia and Chile, may be implicated in the causation of gastric cancer.

High concentrations of nitrates in ground water may result from thick deposits of guano, from certain volcanic rocks, or from prolonged heavy use of organic or artificial fertilizers. Nitrates are also a final product of the oxidation of organic compounds and are therefore associated with organic pollution. Surface or ground waters which receive organic pollution from sewage discharges or on-site sanitation systems may show high nitrate levels. A rising nitrate level in ground water is a warning sign of continuing pollution. Boiling the water will kill any pathogens present, but is likely to increase the nitrate content further.

2.5 REFERENCES AND FURTHER READING

Commins, B.T. (1978). Water is not just H$_2$O, *Nutrition Bulletin*, **4**, 380–393.

Davies, G.N. (1974). *Cost and Benefit of Fluoride in the Prevention of Dental Caries*, Offset Publication No. 9 (Geneva: World Health Organization).

Krishnamachari, K.A.V.R. and Krishnaswamy, K. (1974). An epidemiological study of the syndrome of Genu Valgum among residents of endemic areas for fluorosis in Andhra Pradesh, *Indian Journal of Medical Research*, **62**, 1415–1423.

Miller, D.G. (1980). *Organic Compounds in Drinking Water; An Operational and Economic Assessment*, Technical Report TR149 (London: Water Research Centre).

Panel on Nitrates (1978). *Nitrates: An Environmental Assessment* (Washington DC: National Academy of Sciences).

Shuval, H.I. and Gruner, N. (1977). *Health Effects of Nitrates in Water*, Report No. EPA-600/1-77-030 (Cincinatti: US Environmental Protection Agency).

Van Rensburg. J.F.J. (1981). Health aspects of organic substances in South African waters—opinion and realities. *Water South Africa*, **7**, 139–149.

WHO (1970). *Fluorides and Human Health*, WHO Monograph Series No. 59 (Geneva: World Health Organization).

WHO (1984). *Guidelines for Drinking Water Quality*. (Geneva: World Health Organization).

WHO Regional Ofice for Europe (1979). *Sodium, Chlorides and Conductivity in Drinking Water*, EURO Reports Series No. 2 (Copenhagen: World Health Organization).

Wilkins, J.R., Reiches, N.A. and Kruse, C.W. (1979). Organic chemical contaminants in drinking water and cancer, *American Journal of Epidemiology*, **110**, 420–448.

3

Water Quality and Standards

3.1 DRINKING WATER QUALITY

More stringent control of water contaminants and higher quality standards apply to water intended for human consumption that for other uses. Standards are expressed in terms of the microbiological, the chemical, and the physical characteristics of the water. These are discussed in turn.

Microbiological characteristics

The microbiological quality of drinking waters is typically expressed in terms of the concentration and frequency of occurrence of particular species of bacteria. Polluted water may contain pathogenic (disease-causing) bacteria, viruses, protozoa, or helminths' eggs. The detection and enumeration of all these pathogens on a routine basis is far too complex and, in any case, many of the pathogens will be present only in very small numbers, if at all. It is therefore normal practice to detect and enumerate only what are called 'indicator bacteria'. These are bacteria which are always excreted in large numbers by warm-blooded animals, irrespective of whether they are healthy or sick. The presence of indicator bacteria in water is therefore indicative of faecal contamination of that water. If a sample of water is faecally contaminated it may contain any pathogen which is being excreted by the animal and human population causing the faecal contamination, and faecal contamination suggests the potential presence of pathogens and thus a health hazard.

By convention, the most commonly used indicator bacteria are the coliforms. Water is tested either for the presence of the total coliform group or for the presence of faecal coliforms only. Faecal coliforms, mainly comprising *Escherichia coli*, are a subgroup of the total coliform group and they occur entirely, or almost entirely, in faeces. By contrast, other members of the coliform group can be free-living in nature and therefore their presence in water is not necessarily evidence of faecal contamination. *Escherichia coli* are always present in faeces; most are not pathogenic, although some strains are a major cause of childhood diarrhoea throughout the world.

One must distinguish clearly between the examination of chlorinated and unchlorinated water supplies. If one is examining a chlorinated water supply one knows that, if the chlorination process is working correctly, all coliform organisms will have been killed. Therefore, the presence of any coliform organisms, even in very low concentrations, in a chlorinated water supply is indicative, not necessarily of a health hazard, but certainly of a failure in the chlorination system. If the chlorinated water is being tested at a tap at some distance from the treatment works, then the presence of coliforms may indicate a failure in the chlorination plant but it may also indicate the introduction of contamination at some point in the distribution system. Bacteriological standards for chlorinated water supplies are therefore extremely stringent and tests for the total coliform group may be used. However, it is simpler to test for the chlorine directly (Chapter 6), and it is usually appropriate to test very frequently for chlorine and much less frequently for coliforms.

In testing untreated water supplies, such as those used by most rural people in developing countries, a quite different philosophy must apply. An untreated water, like any other natural water, contains active plant and animal life and will contain many bacteria. Nearly all these bacteria will be of no health significance and their presence in the water is quite irrelevant to the public health worker. The only bacteria which are indicative of a health hazard are the strictly faecal bacteria. In testing an untreated water supply therefore, it is necessary to enumerate bacteria of definite faecal origin. The faecal coliforms (*Escherichia coli*) are most commonly used for this purpose, but other groups of bacteria, such as the faecal streptococci or the clostridia, may be used as alternatives.

The use of ratios of faecal coliforms to faecal streptococci to determine whether the contamination is of mainly human or animal origin is common, especially in the American literature. This has no scientific basis, however, and there is no simple method currently available for distinguishing human from animal faecal contamination in water.

Microbiological analysis is not the only approach. Reliable and more informative results may usually be obtained more quickly and cheaply from a 'sanitary survey' of an untreated water supply to check for defects such as leaks in well linings and pipes and the proximity of latrines, rubbish dumps, and other sources of pollution (WHO, 1985).

Microbiological tests

Faecal bacteria in most waters die off with time, dying faster in warmer water and in sunlight. It is therefore necessary to be able to transport water samples to the laboratory within a few hours of collecting them and to keep them cool and shaded on the way. This can be very difficult in a tropical country with a limited road system and requires careful organization.

Every country should be equipped with at least one, and usually several laboratories able to carry out, rapidly and routinely, tests for the presence of coliforms and faecal coliforms in drinking water supplies. Two main laboratory methods may be used. First, there is the 'most probable number', also called the 'multiple tube', method in which the water under consideration is mixed with a nutrient medium (often MacConkey broth) and is incubated for a particular time (usually 24 hours) at a particular temperature (usually 37°C for total coliforms and 44°C for faecal coliforms). The production of acid (which turns the purple MacConkey broth yellow) and of gas (which is caught in a small inverted tube within the liquid) are confirmation of the presence of coliform bacteria. A number of test tubes are incubated, containing different volumes of the water under examination, and by observing the number of positive reactions (in other words the production of both acid and gas) it is possible to derive from tables the most probable number (MPN) of coliform bacteria in the original water. This method is described in detail in a number of texts (APHA, 1989; Mara, 1975).

The alternative method is the 'membrane filtration' technique in which water is filtered through cellulose membranes having a pore size of only 0.45 μm. These retain all bacteria on their surface and the membrane is then placed on a nutrient medium (broth or agar) and incubated for a given time at a given temperature. Bacteria, of the species or group which are favoured by the particular medium, will grow into visible colonies on the membrane and will take on a particular colour depending on the stains contained in the medium. They may then be counted. Both the most probable number and membrane filtration results are expressed as a concentration of bacteria/100 ml. Membrane filtration is more accurate since it allows the bacteria to be counted directly, but the multiple tube test is cheaper, requires less specialized equipment, and is easier for inexperienced laboratory staff to carry out. A description of the multiple tube test in the form of a simple manual is given by Mara (1986).

Microbiological standards

When considering a chlorinated and treated water supply the question of standards is a relatively straightforward one. The presence of even very low concentrations of coliforms is indicative of a failure in the treatment plant or subsequent pollution of the treated water. Therefore standards are typically very rigorous. The World Health Organisation (WHO, 1984) suggests the following guidelines for treated drinking water:

(i)   water entering the distribution system should contain no coliform organisms;

(ii)   water at the tap should contain no coliforms in 95% of samples taken in any one year and it should never contain more than 3 coliforms/100 ml or any *Escherichia coli*.

For untreated water supplies, and particularly for the water supplies of rural people in the tropics, the question of standards is much more difficult. Untreated water sources are almost invariably contaminated with faecal matter and contain faecal coliforms and other indicator bacteria. It is therefore more important to determine the concentration of indicator bacteria in a sample of water than simply to demonstrate their presence. Table 3.1 gives some typical values. In general, surface-water sources may be expected to be substantially polluted in any area with significant human or animal populations, whereas groundwater sources are often of better quality. However, it would be almost inconceivable to find any untreated water supply in any village in any developing country in which one could not detect faecal coliforms and other faecal bacteria. Therefore, to apply standards as stringent as those laid down above would be to condemn the water supplies used by the great majority of the population of most developing countries.[1]

At best, such a ruling will simply be ignored and thus bring similar regulations into disrepute; at the worst it may force people to abandon improved but lightly contaminated supplies in favour of the only alternative, which may be unimproved and heavily polluted sources of water. For example, there have been cases where over-zealous health officials have closed down shallow tube wells because they were found to contain 50 faecal coliforms/100 ml and have thus forced villagers to use polluted irrigation canals containing 10^4 faecal coliforms/100 ml.

A useful role for bacteriological water quality testing in villages using untreated water supplies is to select between alternative sources

---

[1] WHO (1984) does suggest guidelines for unpiped water supplies — namely less than 10 coliforms and no *Escherichia coli*/100 ml. If applied as standards, we believe that these values are too stringent to be helpful in most circumstances; many water supplies used by small communities and farms in the upland areas of the United Kingdom do not conform to them. We understand that a more flexible approach will be proposed in the forthcoming edition of the WHO *Guidelines for Drinking-water Quality*.

Table 3.1 Some reported concentrations of faecal coliforms in untreated domestic water sources in developing countries

Source	*Escherichia coli* per 100 ml[1]
Gambia:	
Open, hand-dug wells, 15–18 m deep	Up to 100 000
Indonesia:	
Canals in Central Jakarta	3100–3100 000
Kenya:	
Springs	0
Dam	0–2
Waterhole	11–350
Large river	10–100 000
Lesotho:	
Unprotected springs	900
Waterholes	860
Small dams	260
Streams	5000
Protected springs	200
Tap water (springs)	9
Tap water (boreholes)	1
Nigeria:	
Ponds	1300–1900
Open hand-dug wells,	200–580
Tap water (borehole)	Up to 35
Nigeria:	
Ponds	4000 000[2]
Open hand-dug wells	
6–12 m deep	50 000[2]
Stored in home	100[2]
Papua New Guinea:	
Streams	0–10 000
Tanzania:	
Rain water	3
Waterholes	61
Ponds	163
Streams	128
Unprotected springs	20
Protected springs	15
Open wells	343
Protected wells	7
Boreholes	1
Treated tap water	3
Uganda:	
Rivers	500–8000
Streams	2–1000
Unprotected springs	0–2000
Protected springs	0–200
Hand-dug wells	8–200
Boreholes	0–60

[1] When only a single value is given it is a geometric mean.

[2] Total coliforms rather than faecal coliforms.

These figures are not necessarily typical of the domestic water quality in the countries concerned. They are measurements taken from selected sources during specific investigations. It is generally true, however, that people in developing countries who must use surface sources or open wells are often drinking water with >1000 faecal coliforms/100 ml.

Source: From Feachem (1980). Reproduced by permission of the *Lancet*

of water. If a number of alternative sources are tested, it is possible to determine which consistently provides water of the best bacteriological quality — particularly after rain. This source many be used in any improvement or extension of the supply. If the improvement includes protection to the source, such as a well cap or spring box (see Chapter 5), this should considerably improve the quality of the water.

After a supply has been built, it requires regular monitoring. In the United Kingdom it is recommended that water from supplies serving less that 10 000 people should be tested weekly, and larger supplies daily. Few developing countries could attain this frequency, and the intensity of monitoring must be chosen to suit the resources and manpower available (WHO, 1976). Samples should preferably be taken several times a year for testing, certainly from supplies serving over 1000 people. An increase in the level of pollution in a particular supply, or much higher level than found in other, similar supplies, is suggestive of a technical fault allowing undesirable contamination of the water.

A rather different use of bacteriological water testing is in the investigation of outbreaks of potentially water-borne disease. If an outbreak of a potentially water-borne disease (Table 1.2, Category 1) occurs, then it is appropriate to test for faecal bacteria in the various water supplies in the area where the outbreak occurred. This may well help the health authorities by showing whether or not there is a source of water which is substantially contaminated by faeces. In such an investigation, one is interested in finding levels of faecal pollution that are substantially above the general norm for that area. Simply to find a few faecal coliforms in the water supply of a village which has had an outbreak of typhoid is in no way a demonstration that the typhoid outbreak was water-borne.

In conclusion, it is necessary to exercise a good deal of common sense in the use and interpretation of bacteriological water quality standards for untreated waters in developing countries. Standards or goals should be set realistically and national authorities must decide themselves what are reasonable levels to aim for, given the particular environmental and economic circumstances of the country. Bacteriological water testing is expensive and should only be undertaken when practical decisions can be taken on the basis of the results.

Chemical and physical characteristics

Chemical and physical water quality standards are commonly laid down for treated waters. These embrace both pollutants which constitute a health hazard (Chapter 2) and qualities of the water which may lead to its having an unpleasant taste, odour, appearance, or other property likely to discourage its use.

For untreated water supplies, chemical water quality standards are generally inappropriate. However, it is necessary to ensure that there are no particular chemical pollutants which constitute a health hazard in certain areas. For instance, some regions of the world such as parts of India and Tanzania have fluoride levels in their ground water which may cause damage to teeth and bones (Chapter 2).

In addition, certain chemical characteristics of untreated ground waters, particularly the content of iron and manganese, may cause the water to be coloured and to stain clothes during washing. This represents no health hazard but may well cause the water-users to reject this ground water in favour of a more bacteriologically contaminated surface-water source. This is a major problem in Bangladesh, for example. It is necessary to ensure, therefore, that any water supply has properties which are acceptable to the population that will be using it. Acceptable levels of these criteria vary very much from community to community. For instance, certain communities which traditionally have always used brackish water will tolerate levels of salt in the water which would prove quite unacceptable to other communities not accustomed to this taste.

The chemical characteristics of ground water cannot necessarily be assumed to be constant. In Tanzania, for instance, fluoride in water from new boreholes has sometimes been found to build up to unacceptable levels only after several months of use. In other places, salt content may be found to rise in the dry season, so that a well found to give sweet water during the rains may not do so all year round. In general, natural springs can be expected to give water relatively free from dissolved chemicals, by comparison with wells and boreholes; the relatively static ground water in the latter has had more time in the ground in which to absorb minerals.

Typical chemical water quality criteria, taken from the World Health Organisation's guidelines, are set out in Table 3.2. Additional WHO guidelines for certain substances are given in Chapter 2.

3.2 WASTE WATER QUALITY

Introduction

Any community which has a water-borne sewerage system will produce a flow of waste water and excreta, generally known as sewage. This sewage may be of domestic, industrial or agricultural origin or typically, a combination of these.

The composition of sewage is both complex and variable. A typical composition is shown in Figure 3.1. The most fundamental characteristics of a sewage are its suspended solids, its oxygen demand, and it content of pathogenic organisms and toxic chemicals.

Table 3.2 Substances and characteristics affecting the acceptability
of water for domestic supply

Substance or characteristic	Undesirable effect that may be produced	Guideline value
aluminium	discolouration	0.2 mg/l
chloride	taste; corrosion to pipes	250 mg/l (as Cl$^-$)
colour	discolouration	15 true colour units
copper	bitter taste; discolouration; corrosion of steel and galvanized pipes	1.0 mg/l
hardness	scale deposition; taste	500 mg/l (as CaCO$_3$)
hydrogen sulphide	odour, taste	not detectable by consumer
iron	taste; discolouration; deposits and growth of iron bacteria; turbidity	0.3 mg/l
manganese	taste; discolouration; deposits in pipes; turbidity	0.1 mg/l
monodichlorobenzene	odour	3.0 µg/l
1,2 dichlorobenzene	odour	0.3 µg/l
1,4 dichlorobenzene	odour	0.1 µg/l
pH	corrosion	6.5–8.5
phenols	taste	1.0 µg/l
sodium	taste	200 mg/l
sulphate	taste; gastro-intestinal irritation	400 mg/l (as SO$_4^{2-}$)
taste and odour	taste and odour	inoffensive[1]
total dissolved solids	taste	1000 mg/l
turbidity	turbidity	5 nephelometric units[2]
zinc	taste	5.0 mg/l

[1] To most consumers
[2] Preferably less than 1 unit for disinfection efficiency
Source: From WHO (1984)

Suspended solids

Sewage contains a complex range of solids, some in solution and
some in suspension. Suspended solids give rise to the turbidity
(or murkiness) of a sewage and indicate the presence of various
chemical and microbiological pollutants. Suspended solids are
determined by filtering a measured volume of the waste water and
retaining the solids on a filter paper which is then oven-dried and
weighed.

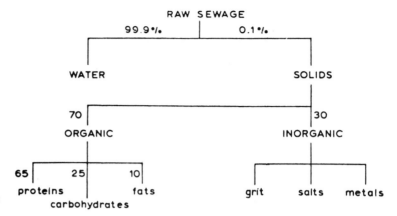

Figure 3.1 The composition of sewage *Source*: From Tebbutt (1992). Reproduced by permission of Pergamon Press

Oxygen demand

The oxygen demand of a sewage is roughly the amount of oxygen required to oxidize the various organic chemicals within it. A raw sewage having an oxygen demand of 300 mg/l requires 300 mg of oxygen to oxidize the organics and any reduced inorganics in one litre, although the latter normally requires only a small proportion of the total. Oxygen demand is thus a gross and indirect measure of the total organic load contained in the sewage. It is a most important measure because the organic content of a sewage is highly complex and defies easy analysis in other ways. Oxygen demands are usually expressed either as a chemical oxygen demand (COD) or a biochemical oxygen demand (BOD). The COD is measured by boiling the sewage with an acid dichromate solution which converts almost all the organics to carbon dioxide and water. The BOD is usually measured by allowing a sample of sewage to stand at 20°C for five days and calculating the amount of oxygen used up during the oxidation of the organics by bacteria. This measure is called the BOD_5 and is related to the amount of *biodegradable* organic matter contained in the sewage. Very approximately:

$$COD = 1.5 \times BOD_5$$

In this book, 'BOD' refers to BOD_5 unless otherwise stated.

Pathogenic micro-organisms

A sewage contains all the types of pathogen being excreted by the contributing population (Table 1.3, Categories I–V). These constitute a health hazard and any decisions about the treatment and disposal of sewage must take this hazard into account.

Pathogenic micro-organisms are not reliably removed from sewage by simply adding a chlorinator to the treatment plant, because of the widely varying chlorine demand of the sewage (Chapter 6), the resistance of some pathogens to chlorine, and the tendency for bacteria to multiply after chlorination. The need to remove pathogens should therefore be borne in mind when selecting the type of treatment works to build.

Toxic chemicals

Sewage with an industrial component may contain heavy metals, phenols and other substances toxic to man and other life. Decisions about treatment and disposal of such sewage must be based on a careful consideration of these toxic effects.

Effluent quality and disposal

Sewage treatment works (see Chapter 10) are generally designed to reduce the concentrations of suspended solids, oxygen demand, pathogens, and toxic chemicals to produce an effluent of a given quality. The desired quality of an effluent from a treatment works depends entirely upon the method of re-use or disposal of the effluent. If the effluent is to be used for irrigation, then certain quite severe standards apply (see below and Chapter 14).

Often the effluent will be discharged into a river, the sea, or less commonly, a lake. The water into which the effluent is discharged, called the 'receiving water', will act to dilute the effluent. The receiving water will also contain oxygen which will help to meet the oxygen demand and so diminish it, and subsequent biological processes will further decrease it over time. Sophisticated methods exist for calculating the impact of an effluent on the quality of a receiving water and, in particular, the changes in the dissolved oxygen content can be calculated.

The quality of the effluent which is required depends on the ratio of effluent flow to receiving water flow (in other words, on the dilution available), on the quality of the receiving water, and on the uses to which the receiving water is put.

Several common features of tropical rivers seriously affect their use as receiving waters for effluent discharges. First, there is often a large difference between maximum and minimum flows. At some times of the year heavy rain in the catchment may cause high flows whereas at other times flows are low or non-existent. At the periods of low flow, the dilution provided by the river may be minimal and the flow may be nearly, or completely, undiluted effluent. In such circumstances, higher effluent quality is required. Second, at warmer temperatures water contains less oxygen and in addition, the rate at which bacteria use up available oxygen is increased. These factors reduce the oxygen available to meet the oxygen demand of the effluent. Third, tropical rivers are often turbid, having a high silt load, and slow flowing. Both these features further reduce the availability of oxygen. Turbid waters inhibit photosynthesis and thus reduce the production of oxygen by algae. Sluggish flow hinders the surface diffusion of oxygen into the water and the dispersal of the effluent.

In the United Kingdom it is common practice to require an effluent to contain not more than 30 mg/l of suspended solids and not more than 20 mg/l of BOD, with a minimum ratio of river flow to effluent flow of 8:1. These standards may act as a guide, but are not universally applicable. For instance, some tropical rivers have a natural BOD over 20 mg/l, and for these it would be more appropriate to specify an effluent BOD not greater than that of the receiving water.

However, in many cases in developing countries the microbiological quality of the effluent is more important, for instance where the inhabitants of downstream villages drink untreated river water. In such instances, it is imperative to incorporate bacterial concentrations (e.g. faecal coliforms/100 ml) into the design calculations.

A more general rule may be that effluent discharges into rivers being used for drinking by downstream villages should not contain a higher concentration of faecal bacteria than the receiving water. This is often a stringent requirement. Conventional treatment of the types used in temperate countries may produce an effluent containing 10^6 to 10^7 faecal bacteria/100 ml, which will substantially pollute a stream containing only 10^2/100 ml. To reduce the effluent to the same quality as the stream may require a chain of four waste-stabilization ponds (see Chapter 10).

3.3 INDUSTRIAL WASTES

The problem

Industrial wastes merit special mention. Factories in developing countries are often located in areas, such as the peripheries of large cities, where the surrounding population uses untreated water from streams and lakes for domestic purposes. Effluent discharge into surface waters then poses a health risk to local communities. Industrial wastes in developing countries may also endanger public health by contaminating irrigated food crops, by contaminating or killing fish which are a source of protein, or by affecting bathing places.

The flow of effluent may be highly seasonal, particularly from the many industries which process agricultural products, and pollution problems will be increased if the peak discharge coincides with the season of low flow in streams and rivers. This is what happened on the Mae Klong River in Thailand, where large amounts of water were used by sugar refineries for cooling and returned untreated to the river. The river water temperature was raised by over 5°C, the COD increased by nearly 200 mg/l and the surface was covered by oil from the cane-crusher bearings. The fish were killed and the population living by the river lost their source of water.

Many different industrial processes produce harmful wastes. Food and drink industries tend to discharge heavy organic pollution and oxygen demand to urban streams used as water sources by low-income groups. Problems are caused by dyes from textile factories, plating wastes from metal-finishing works, and toxic effluents from chemical factories whether they are making industrial or agricultural chemicals, plastics, pharmaceuticals, or cosmetics. Heavy metals, such as mercury and lead, in wastes from mines and factories, are particularly hazardous.

Control

Most developing countries have inadequate environmental control legislation, and no body with the clear responsibility, means, and authority to enforce it adequately. The cost of the pollution control measures required for some industrial processes in Europe and North America may amount to 40% of the capital investment and 15% of the operating costs. Clearly, therefore, there are savings, besides the reduction in its wage bill, which a corporation many make by moving a factory to a developing country, even if that means importing all the raw materials and exporting the finished product. A further difficulty is caused by the ambivalence of decision-makers to large industrial enterprises. Politicians and administrators are unwilling to deter the foreign investment they seek for their country's industrial development. Besides, large foreign companies have the resources and the expertise to evade or suborn the pollution control authority.

It is impossible to lay down standards in advance for all possible pollutants, and the most useful control measure is to empower a single agency, such as the national water authority, to decide and enforce the standards to be applied in each case. The system should be in terms of guidelines rather than standards, so as to be flexible enough to take downstream use into account. It would be wasteful to insist on costly treatment of effluent which is being discharged to a stream which is used for no other purpose. The responsibility for treatment of polluting wastes must rest with the polluter, and it will normally be preferable to apply the special treatment required as the waste leaves the factory, rather than to combine it with other wastes.

Where toxic or hazardous substances are concerned, the legislation should be in such a form that the pollution control agency can define which substances are to be included, that prior consent of the agency is required to discharge them, and that this consent can be conditional and can be revoked. Control will be assisted if a register is kept of all hazardous substances imported or manufactured, including what quantities are involved, and where.

3.4 BATHING WATER QUALITY

Water quality for recreational or religious bathing is not usually a serious problem in developing countries. Epidemiological evidence for a health hazard associated with bathing in polluted water is limited. The great exception to these remarks is the risk of schistosomiasis infection. In areas of schistosomiasis transmission, any freshwater body receiving urine or faecal pollution should be regarded as potentially infected with schistosome cercariae. The only waters likely to be relatively safe are fast-flowing streams and deep water far out from the shore of a lake. The desire to bathe, swim, or just splash about in streams, ditches, ponds, and lakes is responsible for much schistosomiasis transmission—especially among children. In addition, reservoirs near cities (for instance, Nairobi) often become foci for transmission and pose a threat to people yachting, water skiing, or swimming. Water recreation in areas of endemic schistosomiasis must therefore be very carefully controlled and limited to sites where faecal contamination can be completely excluded or where mollusciciding takes place (see Chapters 16 and 17).

3.5 IRRIGATION WATER QUALITY

The quality requirements for irrigation water are severe if applied to sewage, but are not usually hard to satisfy in surface waters which receive sewage, as long as there is adequate dilution. The health hazards and microbiological quality guidelines for agricultural re-use of effluents are discussed in Chapter 14. Chemical quality criteria are discussed below.

The major criteria are the total concentration of dissolved salts, which is measured either by electrical conductivity (EC) or by total dissolved solids (TDS), and the ratio of sodium ions to calcium and magnesium ions, known as the sodium adsorption ratio (SAR). Excessive salinity prevents the proper functioning of plant cells, while excessive sodium reduces the permeability and workability of the soil. Generally, conductivities of less than 100 milliSiemens/m (at 25°C) and SARs of less than 15 are acceptable. However, the effect of salinity and SAR are interdependent; the acceptability for irrigation of various waters is shown in Figure 3.2. The other important chemical parameters relate to boron and dissolved oxygen. Limits for these are set out in Table 3.3.

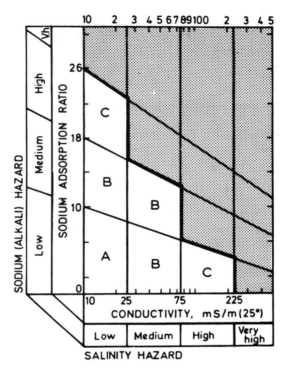

Figure 3.2
Classification of irrigation waters based on their conductivity and sodium and adsorption ratios. Waters in regions A and B are acceptable for almost all irrigation purposes; those in regions C should be avoided wherever possible, and those in the shaded area should not be used at all
Source: From Mara (1976)

Table 3.3 Proposed stream standards for irrigation

Quality parameter	Suggested level of stream standard
Boron[1]	Not more than 1.25 mg/l where there are sensitive crops Not more than 4 mg/l where there are tolerant crops
Dissolved oxygen[1]	Greater than 2 mg/l. A level of 2 mg/l should not occur for more than 8 hours out of any 24-hour period
Faecal coliform density[2]	Not more than 1000/100 ml if the water is to be used for unrestricted irrigation. This standard may be relaxed when the crop is not intended for human consumption

Source: [1]From Pescod (1977). [2]From Chapter 14

3.6 WATER QUALITY AND FISH

Subsistence fishing and commercial fishing are of great importance in many developing countries. Fish provide a vital source of protein for some communities. The most important quality requirement for fish survival is the level of dissolved oxygen in the water; this level is reduced when an effluent exerts an oxygen demand on the receiving

Table 3.4 Proposed stream standards for fishing

Quality parameter	Suggested level of stream standard
CO_2	<12 mg/l
pH	6.5–8.5
NH_3	<1 mg/l
Heavy metals	<1 mg/l
Copper	<0.02 mg/l
Arsenic	<1 mg/l
Lead	<0.1 mg/l
Selenium	<0.1 mg/l
Cyanides	<0.012 mg/l
Phenols	<0.02 mg/l
Dissolved solids	<1000 mg/l
Detergents	<0.2 mg/l
Dissolved oxygen	>2 mg/l
Pesticides	
DDT	<0.002 mg/l
Endrin	<0.004 mg/l
BHC	<0.21 mg/l
Methyl parathion	<0.10 mg/l
Malathion	<0.16 mg/l

Source: From Pescod (1977).

water. However, tropical fish communities largely comprise coarse varieties which can tolerate lower levels of dissolved oxygen, and species which can breathe at the surface. This makes them generally less sensitive to oxygen deprivation than temperate fish.

Information on the effect of various pollutants on fish and river ecology in the tropics is very limited. The standards set out in Table 3.4 have been proposed and are reproduced here as a guide.

3.7 REFERENCES AND FURTHER READING

APHA (1989). *Standard Methods for the Examination of Water and Waste Water*, 17th edition (New York: American Public Health Association).

AWWA (1975). *Simplified Procedures for Water Examination* (Denver: American Water Works Association).

Evison, L.M. and James, A. (1977). Microbiological criteria for tropical water quality, in *Water, Wastes and Health in Hot Climates*. Feachem, R., McGarry, M. and Mara, D. (eds) (London: John Wiley), pp 30–51.

Feachem, R.G. (1980). Bacterial standards for drinking water quality in developing countries, *Lancet*, **2**, 255–256.

GEMS: Global Environmental Monitoring System (1987). *Global Pollution and Health; Results of Health-related Environmental Monitoring* (Geneva: World Health Organization).

Lloyd, B. and Helmer, R. (1991). *Surveillance of Drinking Water Quality in Rural Areas* (London: Longman).

Mara, D.D. (1975). *Bacteriology for Sanitary Engineers* (Edinburgh: Churchill Livingstone).

Mara, D.D. (1976). *Sewage Treatment in Hot Climates* (London: John Wiley).

Mara, D.D. (1986). Bacterial analysis of drinking water, in *Small Water Supplies*. Ross Institute Bulletin No. 10 (London: London School of Hygiene & Tropical Medicine).

Ouano, E.A.R., Lohani, B.N. and Thanh, N.C. (1978) *Water Pollution Control in Developing Countries*. (Bangkok: Asian Institute of Technology).

Pescod, M.B. (1977). Surface water quality criteria for developing countries, in *Water, Wastes and Health in Hot Climates*, Feachem, R., McGarry, M. and Mara, D. (eds) (London: John Wiley), pp 52–77.

Tebbutt, T.H.Y. (1992). *Principles of Water Quality Control*, 4th edition (Oxford: Pergamon Press).

WHO (1976). *Surveillance of Drinking Water Quality*, WHO Monograph Series No. 63 (Geneva: World Health Organization).

WHO (1984). *Guidelines for Drinking-water Quality*, Vol. 1. (Geneva: World Health Organization)

WHO (1985). *Guidelines for Drinking-water Quality*, Vol. 3: Drinking-water quality control in small community supplies. (Geneva: World Health Organization).

Part II

Water Supply

4

Water Supplies in Developing Countries

4.1 THE HAVES AND THE HAVE NOTS

The essential paradox of community water supply in developing countries is that, in one sense, everyone has a water supply while, in another sense, most people have not. Water is essential for life and all human communities must have some kind of water source. It may be dirty, it may be inadequate in volume and it may be several hours' walk away but, nevertheless, some water must be available. However, it we apply any reasonable criterion of adequacy — in terms of the quantity, quality, and availability of water — then most people in developing countries do not have an adequate water supply.

The World Health Organization's figures for 1988 show that, among the urban population of the developing countries, only about 65% have house connections and an additional 20% have access to public taps; about a half of these supplies are intermittent. Of the rural population, only about 60% have access to safe water, and few of these have house connections. The percentages by country are illustrated in Figure 4.1. The figures for many countries are overestimates. Overall, 32% of people in developing countries lack an adequate water supply. These people without adequate water supplies number at least 1200 million.

Table 4.1 shows how the level of coverage has gradually increased over the years. Population growth, particularly in urban areas, means that capacity must be increased simply to prevent the percentage served from falling. Average costs of water provision per capita are given in Table 4.2. Clearly, it will take several more decades and massive investments before water supplies can be provided for all the world's population.

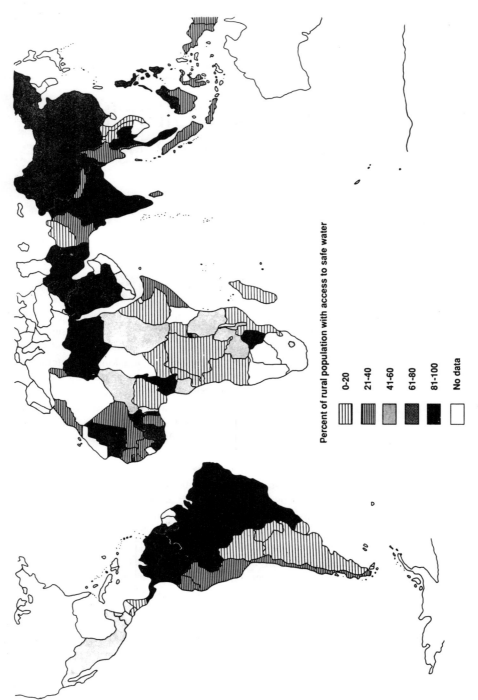

Figure 4.1 Reported percentages of the rural population of various countries having access to safe water in 1988. These figures are optimistic estimates, and some countries have lower coverage than shown.

Table 4.1 Percentage of the population with 'reasonable access' to 'safe water' in developing countries, excluding the People's Republic of China[1]

	1970	1980	1988
Urban	65	73	83
Rural	13	32	55

[1] The 1988 coverage figures for China were: urban 87%, rural 66%.

Table 4.2 Costs of water supply per capita (1988 US$)

Type of service	Construction[1]	Equivalent annual cost[2]	Annual total cost[3]
Urban			
house connection	130	16	23
public tap	45	5	8
Rural	50	6	9

[1] Median values of the costs reported to WHO from 83 countries
[2] Assuming a 20-year lifetime and 10% discount rate
[3] Assuming that 30% of the total is for operation and maintenance

4.2 OPERATION AND MAINTENANCE

However, the construction of new water supplies does not necessarily solve the problem, without the capability to operate and maintain the water supplies which have been built. Many countries have found that construction is relatively easy compared to the task of keeping a large waterworks running continuously, or of servicing hundreds of small village supplies scattered about the countryside. Besides, it is usually easier to obtain development finance for the construction of new supplies than funding for the recurrent expenditure of a maintenance programme. Frequently, not enough money is available to cover operating costs and to carry out running repairs, let alone to carry out necessary preventive maintenance. Operation and maintenance require a long-term commitment of money, staff, and institutions and can be a major drain on the resources of a developing country. Even where the money is available, there is frequently a shortage of technicians, and a lack of viable institutions able to carry out the job.

Water supplies for the larger cities often use sophisticated equipment for which spare parts — and skilled workers to install them — are not easily available. They require chemicals such as chlorine for their operation, which may become difficult or impossible to import. Poor pipe-laying, frequent illegal connections, and soil erosion in the dusty streets of these cities may cause frequent damage to distribution mains, and the public taps are rapidly worn out or broken by heavy use, or even stolen for re-sale. Water rates are extremely difficult to collect from more than a fraction of subscribers, and what can be collected is usually inadequate to cover the full costs of operation and maintenance.

Urban supplies may provide water of doubtful quality, and for only a few hours of the day (Figure 4.2), but only occasionally do they break down completely. In rural areas, however, problems of operation and maintenance have resulted in very high breakdown rates. In a typical developing country, over a quarter of the rural community water supplies may be out of action, with no adequate organization able to repair and operate them. Some countries rely upon community involvement (see below) to maintain supplies, but this method has often failed. Broken-down supplies represent an enormous wastage of investment and, in many countries, the most economical way of bringing good water to more people is to repair the broken supplies rather than build more new ones. Operation and maintenance is the most deficient area in most water supply programmes. *It is very difficult* and requires more efforts, not less, than construction if success is to be achieved.

4.3 APPROPRIATE TECHNOLOGY

It has become fashionable to talk of the need for appropriate technology in the design of water supplies in developing countries. Appropriate technology is the technology which fits the circumstances and is thus appropriate. Technology must be appropriate in terms of cost in order that it can be afforded; it must be appropriate in performance so that it does the job required; and it must be appropriately simple so that it can be operated and maintained.

Unfortunately, there are many cases of inappropriate technology and most of these arise from the unquestioning export of technologies from Europe and North America to the developing countries. Some engineers trained in Europe and America, or trained with a syllabus borrowed from those areas, have come to believe that the water supply practice developed in the affluent countries can and should be applied everywhere. This is incorrect. Good engineering involves the sensitive application of basic principles to a particular problem so that a solution is derived which is genuinely appropriate to the local context.

Engineering should not involve the rigid application of certain standard designs. The excitement and the challenge of the profession comes largely from the degree of flexibility and ingenuity required to produce appropriate solutions for novel problems. Thus the use of appropriate technology is good engineering and sound common sense.

Figure 4.2 An intermittent water supply in Nigeria. In addition to the cost in the time wasted waiting for water, an intermittent water supply puts public health at risk through shortage of water, potential pollution of the mains, and making household storage jars necessary
(Photo: R Feachem)

4.4 BENEFITS

The technology used must be appropriate to the various costs and benefits of improved water supplies. Engineers are accustomed to designing with cost in mind, but as far as water supplies are concerned it is not so common to evaluate benefits or to define exactly what objectives it is hoped to reach. This is largely because in the developed countries a high level of provision of water services has come to be regarded as essential, and to be embodied in codes of practice and legislation. The available resources are generally adequate to provide this level of service, so that it has not been necessary to consider carefully whether the major objective is concerned with health, or convenience, or something else.

However, for the majority of the world's population in the poor villages and urban slums of the tropics, there is no possibility with the available resources of having the same high level of water provision enjoyed by the people of Europe and North America, and so decisions about the level of service and the type of technology are not pre-ordained but are important aspects of the design. The types of water supply improvement with the most benefits, at a given limited cost, should be determined so that the anticipated cost-effectiveness of alternatives can be compared.

Bold and sometimes wildly exaggerated claims have been made for the benefits of water supplies. The major ones occur in three fields: production, health, and the saving of time and energy in the water-collection journey.

Production

It is here that the most unrealistic claims are made. The development of a new water-using industry or agricultural practice may often appear to be dependent on a water supply, but in practice this is not usually a sufficient condition for such development, and in many cases not even a necessary one. For example, many industries prefer to build their own water supplies rather than use the public ones.

The water requirements of domestic use and of large-scale irrigation are difficult to meet in a combined system. Domestic water supplies require relatively small amounts of good quality water, and to supply it in quantities suitable for irrigation would be wasteful in most cases. When domestic water supplies are used for cattle-watering, careful planning is required to avoid overgrazing around water points.

Health

In urban water supplies, water quality is clearly important. Traditional water sources in cities are more liable to faecal pollution, and liable to infect more people, than in rural areas. Many great water-borne *epidemics* have been caused to defective urban water supplies. However it seems that most *endemic* cases of faecal–oral disease (Table 1.2, Category 1) are not water-borne but water-washed. Insofar as they are water-borne, improvements in water quality will reduce their incidence (Figure 4.3).

A more general benefit results from increased availability and accessibility of water if it leads to an increase in the volume of water used for hygiene, as this affects all water-washed diseases (Table 1.2, Categories 1 and 2). Observation of people's behaviour in various rural settings (White *et al.*, 1972; Feachem *et al.*, 1978) suggests that when water is available within about 1 km or within half-an-hour's return journey of the home, water use does not significantly increase when the distance or time is reduced, until it is less than

Figure 4.3 This girl and her family in Sarawak have clean water. If they do not drink other water, they are protected against water-borne transmission of faecal-oral infections (Category 1, Table 1.2) (Photo: J Abcede, WHO)

100 m. When the point is reached where a tap can be provided within each house or yard, water use may increase dramatically from 10–30 l to 30–100 l/person.day (Figure 4.4). Quantity-related health benefits are therefore most likely where traditional water sources are particularly far away, where queuing at the existing water source is particularly time-consuming, or where water can be supplied to each household.

Simple measures to improve water quality have a dramatic effect where Guinea worm is prevalent (see Chapter 1), and improved accessibility may also help to reduce contact with water infected with schistosomiasis (Chapter 17). Improved access to water has also been shown to be associated with a reduction in eye infections (Prost and Négrel, 1989).

It has sometimes been suggested that improved water supplies may decrease a community's resistance to disease epidemics occurring when they break down. With the exception of poliomyelitis, which is best prevented by vaccination (Figure 4.5), there is no evidence for this.

Time and energy savings, and their money value

These are the most immediate and easily measured benefits, and frequently the most appreciated by the population. The magnitude of the time savings depends on conditions prevailing before installing the new water supply. In rural areas it is typical for women to spend about an hour a day collecting water, and in some communities four

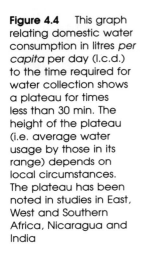

Figure 4.4 This graph relating domestic water consumption in litres *per capita* per day (l.c.d.) to the time required for water collection shows a plateau for times less than 30 min. The height of the plateau (i.e. average water usage by those in its range) depends on local circumstances. The plateau has been noted in studies in East, West and Southern Africa, Nicaragua and India

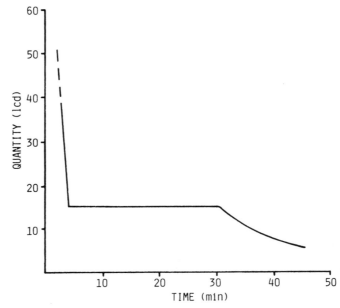

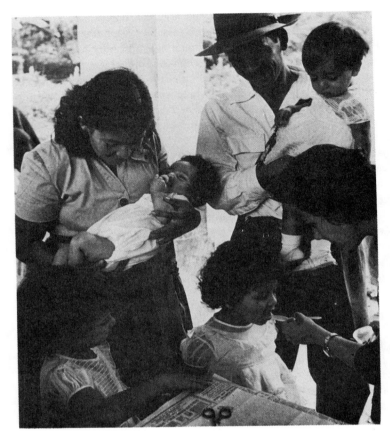

Figure 4.5 Children receive poliovirus vaccine in Colombia. Poliomyelitis is a water-related (Category 1, Table 1.2) and excreta-related (Category I, Table 1.3) infection. It has a low infectious dose and transmits very readily even in affluent communities. It is difficult to control by improved water supply and sanitation and in practice is controlled by vaccination. Hepatitis A and rotavirus infections are also difficult to control by environmental means and may be controlled in future by vaccination (Photo: H Page, WHO)

or five hours a day are required (Figure 4.6). In some urban areas, the time spent collecting water is due not so much to distance as to the long queues which form at water points. It is not usually practicable to predict how the time saved will be spent, but it represents a significant improvement in people's standard of living, and can be regarded as a benefit in itself.

Reductions in the cost of water to the poor are an important, but often neglected benefit of urban water supply schemes. It has been estimated that 20–30% of the urban population of the developing world, mainly the poorest, buy water from vendors, who sell it from tanker trucks, donkey carts, or buckets hung from their shoulders. Typically, these households spend one fifth of their income in this way, though the poorest pay a larger proportion (Cairncross and Kinnear, 1991). In fact, the amount they pay to informal water vendors in many cities is greater than the total revenue of the formal water agency—although it is often the agency's water which is being re-sold! Water vending can also found in some rural communities.

Figure 4.6 Women and children collect water in most communities. In some dry places they spend a large part of their day on this strenuous work. Here children struggle with a heavy water can near Bogota, Colombia (Photo: F Tisnes, Earthscan)

The fact that poor housewives are prepared to pay for water delivered to their door, rather than collect it themselves, shows that they put a money value on their time. For appraisal of the benefits of a water project, a reasonable valuation of this time is the local unskilled wage rate (Churchill *et al.*, 1987). Since this is not far from the value the poor themselves place on their time (Whittington *et al.*, 1989), it can also be used to estimate their willingness to pay for the service.

4.5 COST RECOVERY AND THE PRIVATE SECTOR

At the meeting in New Delhi which marked the end of the International Drinking Water Supply and Sanitation Decade (1981–90), a statement was agreed which included the resolution that,

> A changing role of government is envisaged, from that of provider to that of promoter and facilitator, enabling local public, private and community institutions to deliver services.

The interest in changing the role of governments is linked with a concern to increase the level of cost recovery, for both urban and rural water supplies. Payment of the full cost of water supply by the consumers brings several advantages;

- the rate of water supply construction is not limited by the budgets of governments and aid agencies;
- scarce government funds are less likely to be used in providing luxury supplies for the urban middle class;
- water supplies are more likely to meet people's real needs if they are built only when people are willing to pay for them;
- local bodies can become the principal investors in local water supplies (Briscoe and de Ferranti, 1988), and keep a better eye on how effectively their money is spent than a distant government department.

The last point is more contentious than the others; charging for water does not automatically create well-organized, self-sustaining local institutions, but it does make their creation more feasible.

Many poor people in the Third World already pay substantial amounts to water vendors, so that a payment for a cheaper, more reliable water supply may represent a *saving* to them, not an extra charge. As mentioned above, poor women value their time, and are prepared to pay for a service which saves it for them. Their willingness to pay for a water supply depends largely on the amount of time it saves for them, by comparison with their old source.

Uncontroversial as it might appear, the New Delhi statement also reflected an ideological current in favour of the privatization of government activity which became increasingly prevalent during the 1980s. The privatization discussion is as contentious in developing countries as it is in Europe, and the issues are often over-simplified (Cairncross, 1987). However, a few brief points can be made here.

(a) Public ownership of water supplies is not a question of socialism. A significant proportion of the water supplies in the USA is in public or municipal ownership.
(b) Privatization does not necessarily guarantee greater coverage or better cost recovery. Companies avoid serving poor communities if it is unprofitable, or they may exploit consumers or extract subsidies from governments.
(c) Provision of piped water supply is a 'natural monopoly'; competition can involve a waste of resources, for instance if two pipes run down the same street.
(d) Privatization is not an all-or-nothing affair. Certain functions, such as design and construction, are often contracted out; others could be, such as financing, operation, and revenue collection, while retaining ownership in public hands.
(e) It is not only private companies which could take over some of government's work in the sector; local bodies such as residents' associations and cooperatives also have a role to play.
(f) Promotion and regulation are at least as difficult to do effectively as provision; privatization is not the easy way out for an inefficient government agency, and the New Delhi statement is by no means a mandate for government inaction.
(g) In most developing countries, private enterprise is already active in the water sector, in the shape of borehole drillers, artisan well-diggers, and vendors who sell water at the street corner or door-to-door, often without the knowledge of the formal water agency, and certainly without its promotion or regulation. The first step for government agencies, therefore, is to study this existing activity to see how best to promote it.

4.6 RURAL SUPPLIES AND SELF-HELP

Many governments feel that the task of building, and more especially maintaining, thousands of scattered rural water supplies places an intolerable burden on their scarce resources. They have therefore turned to community involvement and self-help programmes as a way of passing a share of the commitment to the beneficiaries of the supplies. This policy has been supported by the belief that community involvement stimulates a new sense of responsibility and dynamism in the community and leads to more rapid progress in rural development.

There is a great deal of debate about the pros and cons of community involvement and some complex sociological issues are involved. There are many examples of self-help programmes which have failed and two particular causes of failure are worth mentioning.

First, governments typically call upon the villagers to raise money (often 50% of the cost of the supply) or dig trenches and promise to respond with matching funds and technical support. In some cases, the enthusiasm in the villages has been substantial and has far exceeded the government's ability to respond. The government has been left in the embarrassing position of being unable to fulfil its side of the bargain while hundreds of villages wait to receive support.

Second, while village participation in construction (through contributing cash and labour) has been successful in many countries, village participation in operation and maintenance has often been minimal. There are many good reasons why rural communities find it harder to run their supplies than to build them, and poor maintenance is not necessarily attributable to lack of skill or motivation on the part of the villagers. It is more likely to reflect too simplistic a policy, of the form 'if the people help to build them they will look after them', which is not realistic in practice. It is worth noting that villages in Europe and North America do not look after their water supplies on a voluntary basis and would probably find it impossible to do so.

Self-help schemes undoubtedly have a role in many countries. However, they need to be based on a very carefully worked-out partnership between village and government and the responsibilities assigned to the village must be realistic. In general, a village will be unable to raise cash regularly for the operation and maintenance of a public water supply unless it has statutory powers to do so, or a communal income of some kind. Governments cannot escape a substantial commitment of money, staff and institutions to the operation and maintenance of rural water supplies. Roughly speaking, one full-time maintenance worker is required for every 2000 people served, as well as the necessary tools and replacement parts.

Development literature is full of references to the need for a community participation, although there is rarely more than a crude or quite unrealistic analysis of how it may be encouraged, by whom and for whom. It is a false hope to think that all that is required is the placement of community development workers, *animateurs*, or sociologists in close and sensitive contact with the community. Such interventions from outside the community cannot be implemented on a wide scale as they are usually impractical, always expensive in money and staff, and ineffective in the long run when they conflict with the trend of local and national politics. They are an inefficient surrogate for sound local government structures (Feachem, 1980).

Local government requires local power. Local bodies must be able to make and carry out decisions. For this their authority must be recognized by central government and accepted by the public, and they will generally require some financial means. A local council or party committee may decide to build a road before a water supply. This may frustrate the water engineer, but that is what community participation involves. Real community participation takes time. In general, it will be more effective for central government to assist local bodies to take decisions and carry them out (for instance, by training members in accountancy) than to force them into programmes which do not awaken their interest.

4.7 EVALUATION

During the 1980s, increased awareness of the problems of rural water supply programmes, frequently represented by unacceptably high breakdown rates, led to greater consciousness of the need for evaluation. International donor agencies have been especially active in supporting a number of evaluations of water supply programmes, often in the form of epidemiological studies which seek to measure a health benefit, particularly an impact on diarrhoeal diseases in young children.

The experience of these studies has not been very satisfactory for those who funded them. If such a study is carried out with adequate rigour it is likely to cost well over $100 000 and to take a year or more, and it is not guaranteed to detect an impact, even from the most successful programme, or even to produce any meaningful results. One reason for this is that diarrhoea is caused by many different pathogens and has many different transmission routes. It is not possible to tell in practice how each child caught diarrhoea, or even which pathogen is to blame in each case.

Another problem is that the incidence of diarrhoea may be influenced by a multitude of other factors, besides access to a water supply. Examples include the socio-economic status of the household, the education of the mother, access to health care, the prevalence of malaria and the size of the village. Unfortunately, these factors also often happen to be associated with whether or not a household has access to a water supply, and can then cause spurious 'benefits' to appear (or real benefits to disappear) in a study, by a process known to statisticians as 'confounding'. Other problems in such epidemiological studies have been described by Blum and Feachem (1983).

More fundamentally, a health impact study has little diagnostic power for those who run the programme or support it. It does not itself shed light on why a health benefit may not have materialized, or how it could be improved. But the aim of evaluation is precisely to learn from experience and find how projects can be improved in the future. Thus, while health impact studies make a fascinating exercise for researchers, they are of little use as a tool for the operational evaluation of a programme.

A more useful approach is set out in the World Health Organization's Minimum Evaluation Procedure (MEP) for Water and Sanitation Projects (WHO 1983). This approach arose from an understanding of the causal chain which leads from construction of a water supply to any benefits which may result:

$$\text{construction} \longrightarrow \text{functioning} \longrightarrow \text{use} \longrightarrow \text{benefits.}$$

A water supply cannot bestow benefits if it is not used. Nor can it be used if it is not functioning. So the MEP approach is to look first at whether the water supplies are functioning, and whether they are being fully and correctly used. This can be done much more quickly and cheaply than an epidemiological study, and will produce much more useful information for the programme planners.

If an increase in domestic water use is detected, there is a good chance that considerable health benefits will result, as most of the increase is likely to be used for hygiene purposes. If water use is being observed, it may also be possible to collect information about the other major benefit — time saving. A comprehensive guide to evaluation of rural water supply programmes is given by Cairncross *et al*. (1980).

4.8 HYGIENE EDUCATION

Studies of the health impact of water supply and sanitation improvements have generally found that changes in behaviour have played a central role. Some changes may follow automatically from the availability of the new water supply or toilets; an example would be the increase in water consumption in certain cases, as described in Section 4.4 above. However, it is now widely believed that water supply and sanitation programmes should be accompanied by hygiene education, to ensure that the greatest possible health benefits are achieved.

Health education is often badly delivered in developing countries, and there is little scientific evidence of the relative merits of the various possible approaches (Loevinsohn, 1990). However, a few general principles are clear.

(a) For hygiene education to be a genuine component of a water and sanitation project, it should use the field workers who install the facilities (well diggers, latrine builders, etc.), or at least use staff who travel with them to the field.

(b) Participatory methods, involving discussion, are more effective than those in which an audience listens or watches passively.

(c) People are more ready to accept what they hear from neighbours and friends than from outsiders.

(d) Messages should build on existing positive beliefs and practices, rather than try to deny or suppress negative ones.

(e) All messages and media should be pre-tested and the hygiene education programme constantly monitored, to ensure they are appropriate and effective, and produce results.

A number of helpful manuals on the subject are available (WHO, 1987; IRC, 1991).

4.9 REFERENCES AND FURTHER READING

Blum, D. and Feachem, R.G. (1983). Measuring the impact of water supply and sanitation investments on diarrhoeal diseases: problems of method-ology. *International Journal of Epidemiology*, **12**, 357–365.

Briscoe J. and de Ferranti, D. (1988). *Water for Rural Communities: Helping People Help Themselves* (Washington DC: World Bank).

Cairncross, S. (1987). The private sector and water supply in developing countries: partnership or profiteering? *Health Policy and Planning*, **2**, 180–182.

Cairncross, S. (1990). Health impacts in developing countries: new evidence and new prospects. *Journal of the Institution of Water and Environmental Management*, **4**, 571–577

Cairncross, S., Carruthers, I., Curtis, D., Feachem, R., Bradley, D. and Baldwin, G. (1980). *Evaluation for Village Water Supply Planning* (London: John Wiley).

Cairncross, S. and Kinnear, J. (1991). Water vending in urban Sudan. *Water Resources Development*, **7**, 267–273.

Churchill, A., de Ferranti, D., Roche, R., Tager, C., Walters A. and Yazer, A. (1987). *Rural Water Supply and Sanitation: Time for a Change*. World Bank Discussion Paper No. 18 (Washington DC: World Bank).

Feachem, R.G. (1980). Community participation in appropriate water supply and sanitation technologies: the mythology for the Decade, *Proceedings of the Royal Society of London B*, **209**, 15–29.

Feachem, R.G., Burns, E., Cairncross, S., Cronin, A., Cross, P., Curtis, D., Khan, M.K., Lamb, D. and Southall, H. (1978). *Water Health and Devel-opment: An Interdisciplinary Evaluation* (London: Tri-Med Books).

IRC (1991). *Just Stir Gently; the Way to Mix Hygiene Education with Water Supply and Sanitation*, IRC Technical Paper No. 29 (The Hague: IRC International Water and Sanitation Centre).

Loevinsohn, B.P. (1990). Health education interventions in developing countries: a methodological review of published articles. *International Journal of Epidemiology*, **19**, 788–794.

Prost, A. and Négrel, A.D. (1989). Water, trachoma and conjunctivitis. *Bulletin of the WHO*, **67**, 9–18.

White, G.F., Bradley, D.J. and White, A.U. (1972). *Drawers of Water: Domestic Water Use in East Africa* (Chicago: University of Chicago Press).

Whittington, D., Mu, X. and Roche, R. (1989). *The Value of Time Spent on Collecting Water; Some Estimates for Ukunda, Kenya.* Report INU 46 (Washington DC: World Bank).

WHO (1983). *Minimum Evaluation Procedure (MEP) for Water Supply and Sanitation Projects* (Geneva: World Health Organization).

WHO (1987). *Communication: a Guide for Managers of National Diarrhoeal Disease Control Programmes* (Geneva: Diarrhoeal Diseases Control Programme, World Health Organisation).

5

Rural Water Supply

5.1 BREAKDOWNS AND TECHNOLOGY

At least one in four of the rural water supplies in most developing countries is out of order. Very often the reason why a rural water supply has not been repaired is connected with the organizational problems mentioned in Chapter 4, but there are usually technical reasons why it broke down in the first place. These technical reasons may at first sight appear to be problems of poor standards in workmanship due to a shortage of skilled and supervisory staff. But such problems are almost inevitable in most developing countries, and the engineer who blames high breakdown rates on them is rather like a workman blaming his tools. Good engineering requires designs which can be made to work, *with the labour and materials currently available*.

For instance, the engineer cannot usually rely on high standards of pipe-laying to regular slopes. The slopes on pipelines should be large enough to allow for air locks due to uneven gradient, and for errors in the original survey. This can be turned into an advantage, because if it means the minimum permitted gradient for gravity-fed pipeline is, say, 1 in 50, then the maximum possible size of pipe required for a given flow is only 60% larger than would be required for a very steep slope of 1 in 5. For a given flow capacity, therefore, the pipe required for any likely slope can be chosen from a range of only two or three consecutive sizes. This means that pipe sizes can be chosen from a table like Table 5.1 by a technician who would find a Hazen–Williams pipe flow chart confusing.

Design for maintenance

If it is important for water supply technology to be chosen so that it can be made to work under the existing construction conditions, it is even more important that it should continue to work under the prevailing maintenance conditions. Water treatment plant, for example, generally requires a level of attention and skill in operation quite unattainable in a small community. Since there is

Table 5.1   Diameters for gravity pipelines (mm)

Flow (l/s)	Steel		Polythene		Bamboo		PVC	
	Flat	Steep	Flat	Steep	Flat	Steep	Flat	Steep
0.10	19	19	12	12	25	19	19	12
0.15	25	19	19	12	32	25	19	19
0.20	25	19	19	12	32	25	25	19
0.30	32	25	25	19	32	25	25	19
0.40	32	25	25	19	37	32	25	25
0.60	37	32	32	25	50	32	32	25
0.80	50	32	32	25	50	37	37	32
1.00	50	37	37	32	62	50	37	32
1.50	62	50	50	32	76	50	50	37
2.00	62	50	50	37	76	62	50	37
3.00	62	50	62	50	76	62	62	50

'Flat' is < 1:15
'Steep' is > 1:15

little point in installing water treatment facilities if they will not be reliably operated, it is almost always preferable to find a source of good-quality water and protect it from pollution, rather than to take water from a doubtful source and treat it. Pumps, too, of any kind, frequently break down or fall into disuse in rural areas. Motorized pumps, especially, should only be installed where adequate arrangements have been made to pay for their running costs (see Chapter 4).

It is best to try to build a 'fail-safe' character into rural water supplies so that one small fault is not likely to put the whole system out of action. For example, a ring main is preferable to a 'dendritic' distribution system (Figure 5.1), so that if a pipe is broken the whole community is not necessarily deprived of water. Again, a series of hand pumps on tube wells may have the advantage over a piped supply from a distant water source, because if one pump breaks down the villagers can continue to use the others until it is repaired.

As a general rule the cheaper and simpler the technology, the less maintenance it requires, the more reliable it is in practice, and the easier to repair under village conditions. The major exception is in the choice of hand pumps, where the more robust and reliable pumps are often more expensive and more difficult to repair; here the choice depends not only on technical considerations but also on factors such as whether maintenance is to be carried out by a village caretaker, a local mechanic, or a mobile team. Village Level Operation and Maintenance, known as VLOM, is generally preferable.

5.2 SOURCES OF WATER

Because of the unreliability of treatment plant under most rural conditions, the best sources of water are those which do not need

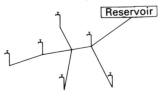

(a) Dendritic system

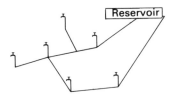

Figure 5.1 Alternative village water supply distribution systems

(b) Ring main

treatment. Rainwater collected from a metal or asbestos cement roof is relatively pure, and is of course available close to the users if the roof is theirs. However, many rural houses are roofed with other materials, such as thatch, and rainfall patterns may require large and expensive storage tanks to guarantee a supply all year round.

Surface water may be readily available and easy to abstract, but is typically very polluted (Table 3.1). In some sparsely populated upland areas, streams may be of a quality good enough for domestic use, but in most regions streams, lakes, and ponds are subject to substantial faecal pollution.

Where it can be extracted with reasonable ease, ground water is normally preferable to surface water because it is purified by the filtering action of the soil through which it flows. Nevertheless, ground water in some areas may contain iron, manganese, salt, fluoride, or other substances which make its use undesirable or unpleasant, and the use of a surface source—a river, lake, or dam—may be unavoidable. Even in these cases, a well beside the surface source usually gives fresh water and is to be preferred. Where it is not ·possible to locate a reliable year-round source of water within the village, a more distant source may be supplemented by a 'wet-season well' which, although it may not be in use in the dry season, at least supplies water during the rains, which is usually the period of greatest disease incidence and peak labour demand. Sufficient quantities of ground water to supply a rural community may be collected from a spring or extracted from a well of some kind.

Protected springs

Springs, where they exist and have a reliable flow, can make ideal sources of water for a community water supply. No pumping is required to extract water from them, and all that is usually necessary to obtain water of good quality is to collect it and protect it from pollution. This is done by building a box of brick, masonry, or concrete around the spring so that water flows directly out of the box into a pipe without ever being exposed to pollution from outside (Figure 5.2).

The water emerging at a spring has generally been forced to the surface by an impervious layer of soil or rock; a layer through which water cannot pass. In excavating the foundations for a spring box it is important to avoid digging through this layer, or the water may seep downwards so that the spring disappears or moves downhill.

The point where the water emerges, known as the 'eye' of the spring, should be covered with carefully selected sand or gravel. If this material is too coarse, the spring water may erode the soil behind it, but it should not be finer than the existing soil behind it or it may block the flow. If there is a danger that it too could be washed away, then still coarser gravel should be placed in front of it, with the gravel progressively increasing in size to the stones of the spring box wall. A spring may sometimes flow very strongly for brief periods after rain, and the whole structure must be sound enough to resist erosion.

Fine sediment is suspended in the water from most springs. A spring box therefore should be built so as to prevent this sediment from settling over the eye of the spring and blocking its flow. This is best done by ensuring that the overflow pipe is not above the eye (Figure 5.2). It is also important for the spring box to have a removable cover, so that it can be cleaned out from time to time. Alternatively, one or more small springs may be connected to a single 'silt trap' (Figure 5.3), where the silt is allowed to accumulate and is periodically cleaned out.

Care is required to prevent surface water from running into the spring box and polluting the water in it. Puddled clay should be used to backfill behind the box to seal the ground against infiltration. The top of the spring box should be at least 300 mm above the ground,

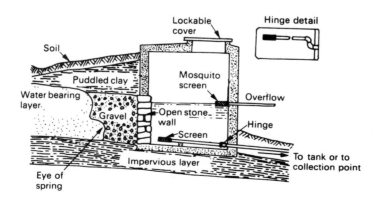

Figure 5.2   A spring box. The inset detail shows a hinge made with two flexible pipe bends, enabling the screen to be lifted above the water for cleaning

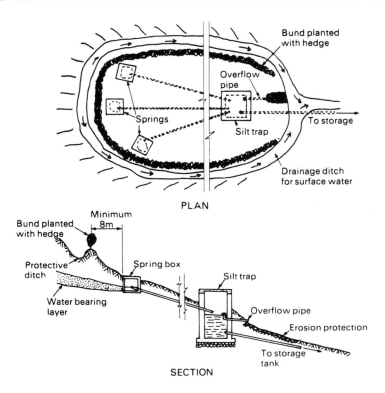

Figure 5.3 Three protected springs connected to a silt trap

and the access hole should have a lip around it and a cover which is not easily removed. In addition, a ditch may be dug on the uphill side of the spring, and the excavated soil thrown up into a bank or 'bund' (Figure 5.3) to divert surface water. Finally, a fence or prickly hedge planted on the bund will help to keep people and animals away.

Wells

Wells can be sunk in a wide variety of ways. The basic methods most suitable for small community water supplies are illustrated schematically in Figure 5.4, and listed below.

(1) *The driven tube well*, in which a specially perforated or slotted tube known as a 'well point' is driven into the ground. The well point is re-usable, but is expensive and normally lasts only about 5 years. Well points can be made locally from galvanized iron pipe (FAO, 1977), but these are more liable to clogging and corrosion than commercial well points.
(2) *The bored tube well*, which can be sunk by hand to depths up to 40 m with an 'auger', a simple tool twisted by hand to drive it into the ground.

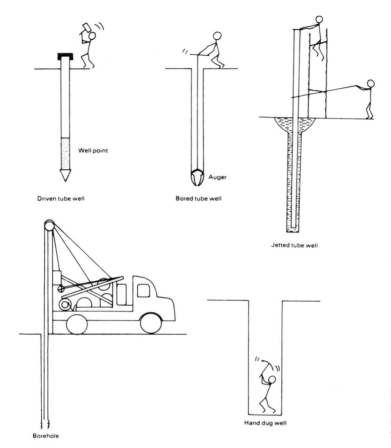

Well point

Driven tube well   Bored tube well

Auger

Jetted tube well

Borehole

Hand dug well

Figure 5.4 Schematic illustration of five basic methods of ground water extraction

(3) *The jetted tube well,* in which a pipe is sunk into soft ground while the soil is loosened and removed by water pumped down (or up) the pipe while the surrounding hole is kept full of water. Figure 5.5 illustrates the simplest method of jetting, known as the 'palm and sludger' method; the pipe is moved up and down by a lever, and a person's hand or a wooden flap on the top of the pipe is used as a valve to pump water upwards, closing as the pipe rises and opening as it falls.

(4) *The hand-dug well,* the most common method of extracting water from the ground. It can be dangerous to build unless the necessary skills are available locally; but, if they are, it can be constructed cheaply with local equipment and materials. The hand-dug well has the very important advantage that water can be drawn from it by bucket and rope if a pump cannot be afforded, or if the pump breaks down.

(5) Various methods of jetting, punching, or drilling a tube well or borehole require a special *drilling rig* which may be trailer- or truck-mounted for use in rural areas. The rigs are expensive, but

many of these methods have equivalents using hand power, or a small motor and pump (FAO, 1977).

The construction of the various types of well is described in several technical manuals (FAO, 1977; Watt and Wood, 1977; Brush, 1979; Blankwaardt, 1984; IDRC, 1981) and so is not discussed here.

Figure 5.5 The palm-and-sludger method in use in Bangladesh (Photo: HEED, Dhaka Bangladesh)

Pollution of open wells

Tube wells and boreholes are protected from pollution by a concrete slab, at least 2 m across, used as a base for the pump (see Figure 5.12). Open hand-dug wells, however, are more liable to pollution. An open well can be polluted by any of the following means, but only the first two normally affect tube wells.

(1) *Polluted ground water*   This can result from location of the well too close to pit latrines, soakaways, or refuse dumps, whose influence may extend to about 10 m in a typical soil. In fissured strata such as limestone and fractured rock, water may flow in underground streams rather than seeping through the soil, and so carry faecal pollution much longer distances (see Section 8.4).

(2) *Seepage water from the surface*   This may enter through the top few metres of the well lining if it is not sufficiently watertight near the surface (Figure 5.6).

(3) *The vessels used for drawing water*   However often these may be rinsed out, they can cause some pollution of the well (Figure 5.7). An improvement can be achieved by having a

Figure 5.6   This well in Burkina Faso is contaminated by seepage of the pool of water around the well head into the well. The pool is contaminated by pig faeces
(Photo: R. Witlin, World Bank)

Figure 5.7 This well is polluted by dirt on the tins and buckets that are lowered into it and also by seepage of spilt water and rainwater

bucket permanently hanging in the well, probably from a windlass, so that it is never taken home and never put on the ground (Figure 5.8). If the bucket is made of collapsible rubber, it is less likely to be put on the ground or stolen. Pollution can only be completely avoided by sealing the well and installing a pump, though this may cost as much as the original construction of the well.

(4) *Rubbish thrown down the well* The chance of this may be reduced by preventing children from playing near the well, but the only certain way to prevent it is to fit a permanent cover over the well and install a pump.

(5) *Surface water* This may be washed straight down the well, especially if the ground surface has sunk, as is often the case when the well does not have an adequate lining. It can be prevented by building a headwall (Figure 5.9), which will also help to prevent animals and people from falling into the well.

(6) *Spilt water* If there is no headwall, or if people stand on the headwall to draw water, water which has splashed against their feet can fall back into the well.

Avoidance of pollution by spilt water is particularly important in regions of West Africa where Guinea worm (*Dracunculus medinensis*) is endemic (see Chapter 1 and Appendix D). Its cyclopoid intermediate hosts are not often found in very deep, narrow wells; but in a shallow well, water splashed off the legs of infected people drawing water, or washing with it, could maintain transmission of the disease.

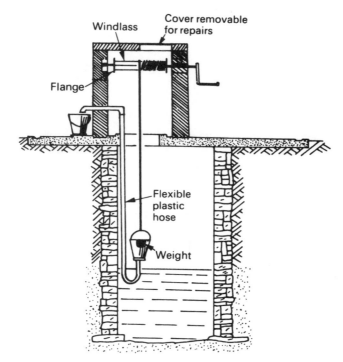

Figure 5.8 One method of protecting a well from pollution. Methods of this kind have not yet been adequately tested in the field

Figure 5.9 A well with a headwall and rollers in Nigeria. This arrangement prevents surface and spilt water running into the well and also prevents Guinea worm transmission at the well (Photo: R Feachem)

Because of this, and because of the importance of surface and spilt water in causing well pollution, the most important single improvement which can be made to an existing well is the construction of a well head consisting of a headwall and drainage apron to take spilt water to a soakaway. This single measure can completely prevent Guinea worm transmission at a well, and considerably reduce other health risks. It can only be done on its own in solid ground where there is no danger of the well shaft collapsing. Where the ground is unstable, it is necessary to build a well lining first.

The headwall can be fitted with rollers, a pulley, or a windlass to help people to pull up the bucket (Figures 5.8 and 5.9). Better protection from pollution can be gained by covering the well with a concrete slab and fitting a hand pump (Figure 5.10). This increases the cost of the well, and a pump should not be installed in any water supply unless arrangements have been made to maintain it.

A well itself requires maintenance too. Dust, rubbish, and dead animals can accumulate remarkably quickly in the bottom of an open well. Apart from polluting the water, the accumulation of rubbish or wind-blown dust in a well may be sufficient to reduce its depth or block it up. Ideally, any open well should be cleaned once a year in the dry season when the water level is low, and then heavily disinfected before being put back into service.

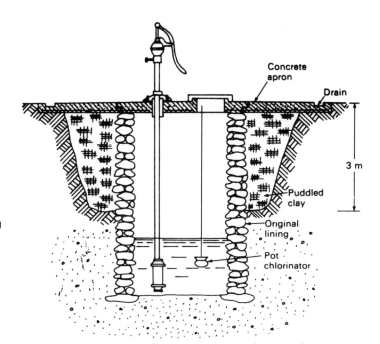

Figure 5.10 Improving an existing well by adding a hand pump, a cover slab, an apron and drain, and a puddled clay barrier against seepage of surface water
Source: After Wagner and Lanoix (1959)

5.3 RAISING WATER

Methods of lifting water are numerous and varied. The simplest mechanisms are often the cheapest, and can more easily be made and repaired with local materials. However, they are sometimes less durable, and usually require more maintenance by the local community. The main methods are described below, roughly in order of increasing complexity and cost. Which of these is most appropriate will depend on the local conditions, the funds available, and the probability of regular maintenance in the future.

Hand power

The first decision to make is whether to use hand power for raising water. Hand power is suitable for a supply where water is drawn straight from the source, such as a well, and the person drawing water operates the device. If water is to be pumped to a storage tank some other type of power will have to be used, such as wind, diesel, or electricity, unless an institutional framework (school, hospital, commune, etc.) exists to organize the work of pumping by hand.

The simplest method of raising water is a bucket of some kind on the end of a rope. It is best to use rollers (Figure 5.9), a windlass (Figure 5.8) or a shaduf (Figure 5.11) so that people do not have to lean over the well headwall to raise the bucket. If the shaduf is designed to balance with the empty bucket in the air, this will help to prevent the bucket from being put down on dirty ground.

However, these devices are not suitable for tube wells, or for very deep hand-dug wells, and for these a hand pump may be required. A hand pump also enables the water to be protected from pollution until it enters the user's bucket, but it cannot usually be made in the village, and may have to be imported. It also requires regular maintenance (Figures 5.12 and 5.13). Most hand pumps employ a piston with leather washers which moves up and down inside a cylinder, rather like a bicycle pump, and the washers must be regularly replaced. Simple hand pumps can be made locally from wood and plastic, but they tend to wear out quickly.

Hand pumps are of two kinds. Shallow-well pumps are cheaper and easier to maintain because the pumping mechanism is above ground level, but they can only work when the water level is less than 8 m deep. In deep-well pumps, the pumping mechanism is immersed in the water at the bottom of the well (Figure 5.12). The 'open cylinder' type of deep-well pump has a riser pipe larger than the cylinder, so that the piston can be pulled up for maintenance using the pump rod, without having to winch up the heavy riser pipe full of water. This makes maintenance by villagers much easier, as the leather washers may have to be replaced as often as four times a year.

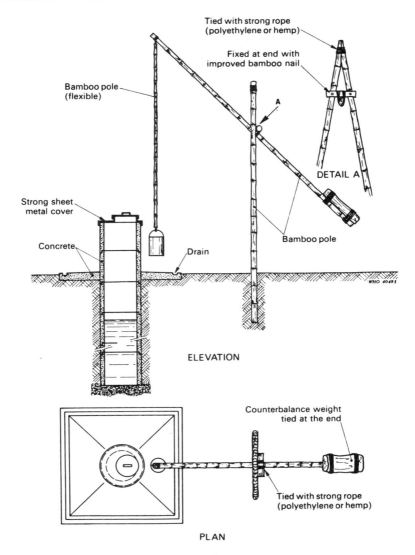

Figure 5.11   A shaduf used over a hand-dug well
Source: From Rajagopalan and Shiffman (1974)

When choosing a hand pump, the following criteria should be borne in mind:

(i)   the pump should be as simple as possible and easy to repair;

(ii) the maintenance required must be easy to carry out and preferably not required too frequently;

(iii) manufacture should not present major quality control problems, and should preferably be undertaken in the country where the pump is to be installed;

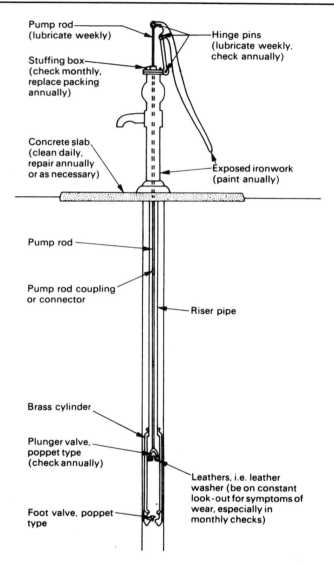

Pump rod
(lubricate weekly)

Hinge pins
(lubricate weekly,
check annually)

Stuffing box
(check monthly,
replace packing
annually)

Concrete slab
(clean daily,
repair annually
or as necessary)

Exposed ironwork
(paint anually)

Pump rod

Pump rod coupling
or connector

Riser pipe

Brass cylinder

Plunger valve,
poppet type
(check annually)

Leathers, i.e. leather
washer (be on constant
look-out for symptoms of
wear, especially in
monthly checks)

Foot valve, poppet
type

Figure 5.12
Maintenance points on
a simple hand pump
Source: From Pacey
(1977)

(iv)   the pump must be reliable;

(v)   the pump should be resistant to abuse, vandalism, and theft
of parts;

(vi)   the pump should be acceptable to users, easy to use, and
produce water at a reasonable rate;

(vii)   the pump should be suitable for the local hydrogeological
conditions—depth of water table, corrosiveness of ground
water, etc;

(viii)   the pump should be accompanied by clearly illustrated installation and maintenance instructions and a basic tool kit;

Figure 5.13 The UNICEF all-purpose hand-pump tool

(xi)   the price should be as low as possible, consistent with the other criteria being satisfied.

Three brands of hand pump which meet some of these criteria are illustrated in Figures 5.14–16. A comprehensive guide to handpump selection is given by Arlosoroff *et al.* (1987), who developed the concept of Village Level Operation and Maintenance (VLOM).

Natural sources of power

Wind power may also be used for raising water, with the advantage that wind is free. However, a windmill is necessary to harness wind power, and windmills are usually rather expensive. A large and expensive storage tank is also necessary to ensure a reasonably reliable supply over windless periods. Even in a quite windy region, storage capacity for seven days' water may be required. Alternatively, a wind pump may be installed which is designed to be operated as a hand pump when there is no wind.

Two other types of pump, the hydraulic ram and the solar pump, use naturally occurring sources of energy. A hydraulic ram uses the energy of flow of a large volume of water, to pump a small proportion of that volume. It therefore requires a much larger flow of water of suitable quality than would be necessary for the community's needs alone. It also requires careful adjustment (Watt, 1975). Solar pumps are suitable for arid areas, they can pump as much as 10 l/s, but they involve sophisticated technology. Several have been installed, with mixed results, in rural areas of francophone West Africa and in Somalia.

Motor pumps

Pumps may also be driven by diesel or electric motors. Electric motors need less maintenance and are usually more reliable than diesel engines, so that they are preferable where electricity is available. Unfortunately, electricity supplies themselves are not always reliable in rural areas.

The simplest motorized well pump works like a hand pump, but a mechanism called a 'forcehead' is used at the top to turn the motor's rotation into the vertical motion necessary for pumping, which is

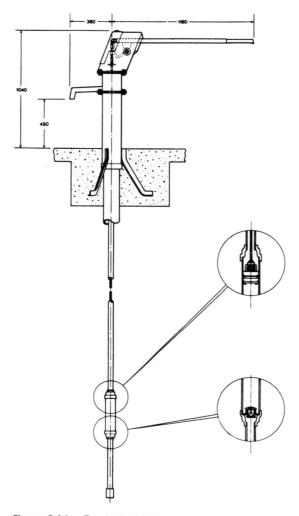

Figure 5.14   The India Mark II pump: a hand-operated deep-well lift pump. It is manufactured in several developing countries (dimensions are in millimetres)
(Photo and drawing: Consumers' Association)

transmitted down the well by a steel rod. In general, it is best to have the motor above the ground where it is accessible for maintenance but to keep the pump mechanism below the water level to avoid the need for 'priming'. Ejector or 'jet' pumps are particularly suitable for boreholes, as both motor and pump are above ground and there are no moving parts down the hole.

There are particular points to note when choosing pumps for village water supplies. Spare parts should be readily available; this can be helped by standardization on one or two basic designs. Motors should be simple to maintain, and be able to run with locally

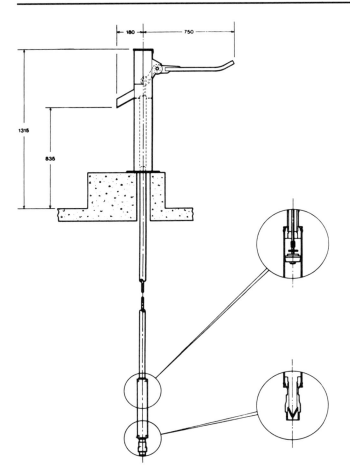

Figure 5.15   The Consallen LD5 pump: a hand-operated deep-well lift pump (dimensions are in millimetres)
(Photo and drawing: Consumers' Association)

available lubricants and local fuel or electricity. The water pumped for a village water supply may be quite silty; if so, a pump should not be liable to wear out too quickly under these conditions. Finally, in view of the difficulty of assessing village demand, it will help if the pump cylinder or, on a centrifugal pump, the impeller, can be changed for a larger one if necessary.

5.4 STORAGE

When designing storage tanks for village water supplies, it is often tempting to use familiar materials such as reinforced concrete or

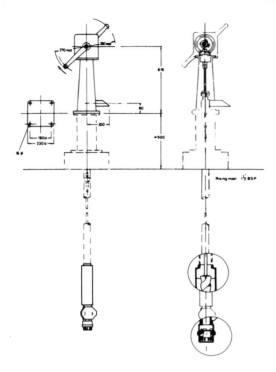

Figure 5.16   A Mono ES30 pump: a hand-operated, rotary helical-screw pump. It is several times more expensive than those shown in Figures 5.14 and 5.15 (dimensions are in millimetres) (Drawing: Consumers' Association. Photo: Mono Pumps Ltd.)

corrugated steel sheet, as these are easier to make reliably watertight. But a few small leaks in a tank above ground may not be serious in village circumstances, and perfectly adequate tanks may be built of local building materials such as brick or masonry, especially if galvanized wire is laid between courses to give the walls horizontal reinforcement. Watt (1978) describes simple methods for building water tanks by plastering cement mortar on to reinforcement of chicken mesh and steel wire (ferrocement).

Small earth dams, too, do not need to be of sophisticated design, being made with clay core walls and rip-rap, although certain basic safety requirements are of course necessary. A good account of small dam construction is given by Wagner and Lanoix (1959).

Care should be taken to prevent tanks and reservoirs from becoming breeding places for malaria mosquitos, especially in seasonally arid areas where malaria transmission decreases during the dry season. The creation of any permanent water surfaces accessible to mosquitoes may promote their breeding in the dry season, unless special precautions are taken. Storage tanks should therefore be covered, ventilation pipes screened with mosquito-proof mesh, and steps taken to avoid the creation of breeding sites downstream from the overflow. Ponds and dams may need special measures to prevent their banks becoming overgrown with weeds, for instance by paving them at water level (compare Figure 10.5).

5.5 TREATMENT

Unfortunately, there is no such thing as a simple and reliable water treatment process suitable for small community water supplies. Therefore it is preferable to choose a source of naturally pure water, and then to collect that water and protect it from pollution so that treatment is unnecessary. Treatment of village water supplies should only be considered if it can be afforded and reliably operated in the future.

Storage

The simplest method of treating water is to store it in a covered tank. Some treatment may be obtained by careful design of storage tanks to ensure a slow and even movement of water from the inlet to the outlet, as in a sedimentation tank. This will permit some silt to settle out, and allow time for some pathogens to die off. If water is stored for at least forty-eight hours, for instance, any schistosome cercariae in it will become non-infective before they leave the tank.

Sedimentation

For larger communities it may be useful to build a small sedimentation tank like that shown in Figure 5.17, although it will not usually be possible to arrange for coagulant chemicals to be added to the water to assist the sedimentation. Sedimentation does not remove many of the harmful organisms from polluted water, but it helps to clarify water for treatment by filtration or chlorination.

Filtration and chlorination

Filtration and chlorination are discussed in Chapter 6, as they are not usually suitable for village conditions. If filtration is unavoidable, it should be by slow sand filters.

One method of chlorination can be used in village wells. It involves a pot containing a mixture of coarse sand and bleaching powder, which is hung underwater in a well (see Figure 5.10). Figure 5.18 shows two types of pot chlorinator.

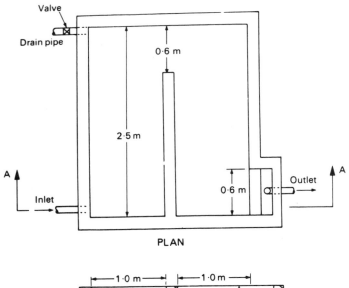

PLAN

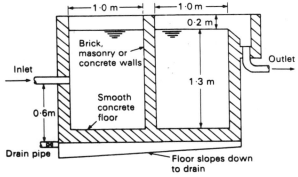

SECTION A-A

a

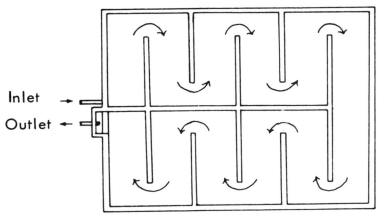

Inlet

Outlet

b

Figure 5.17 (a) A simple sedimentation tank for flows of up to 2000 l/h. (b) A method of combining six small sedimentation tanks to take a flow six times the capacity of each

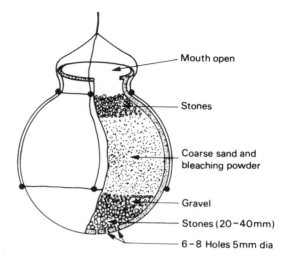

Mouth open

Stones

Coarse sand and bleaching powder

Gravel

Stones (20-40mm)

6-8 Holes 5mm dia

(a) Single pot system

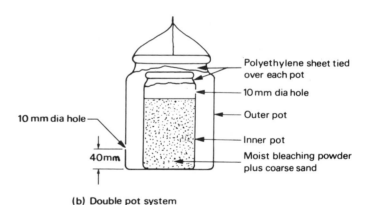

Polyethylene sheet tied over each pot

10 mm dia hole

Outer pot

Inner pot

Moist bleaching powder plus coarse sand

10 mm dia hole

40mm

Figure 5.18 Pot chlorinators for disinfecting wells. Two alternative designs

(b) Double pot system

The double pot is suitable for a well serving up to twenty people, and needs to be refilled with 1 kg of bleach and 2 kg of sand every three weeks. The single pot will serve up to sixty people if it contains 50% more bleach and sand, but it requires replenishing every two weeks. The trouble with these pots is that they tend to make the water taste unpleasant for the first few days after refilling. There is no point at all in using a water disinfection process if it drives people to use water of worse quality, or if it is not reliably operated. Nevertheless, chlorination of a rural water source may be a worthwhile temporary measure during an epidemic which is suspected to be water-borne (Figure 5.19).

Figure 5.19   Emergency well chlorination in Burma
(Photo: H Page, WHO)

Removal of minerals and salts

In a few areas, heavy concentrations of dissolved iron and manganese
in the ground water can give it an unpleasant taste, and give a
brownish colour to food and clothes. These chemicals can be a
serious nuisance, and may even prevent people from using the water.
If so, they can often be removed by aeration, for instance when the
water falls into a storage tank from the inlet. Aeration causes the
iron and manganese to become insoluble so that they form a fine
dark sediment which is more easily removed.

Figure 5.20 shows a simple unit for the aeration of water, which
will also remove the iron and manganese sediment produced. It is
made of four cylinders, the top three of which each have a mesh
or sieve in the base, and ventilation slots in the side. The top two
have a layer of stones 150 mm deep, and the third has a 50 mm
layer of small stones, covered with 300 mm of coarse sand. The
unit stands on a solid brick or concrete platform. Water is sprayed
over the stones at the top and is collected in the bottom cylinder, to
be withdrawn through a tap at the bottom. The water is exposed to

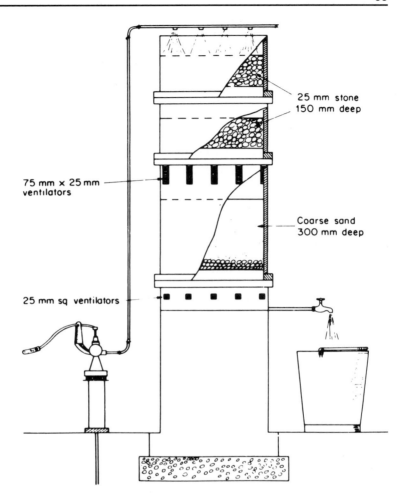

25 mm stone
150 mm deep

75 mm x 25 mm
ventilators

Coarse sand
300 mm deep

25 mm sq ventilators

Figure 5.20 A hand-
operated unit for
iron and manganese
removal
Source: From Pickford
(1977)

the air as it trickles down through the stones, and the sediment is
deposited on the sand lower down. The sand requires replacement
roughly once a month.

Other chemicals in water, particularly salt, fluorides, and nitrates,
are less easily removed under village conditions. The simplest
fluoride-removal process, for example, requires the regular addition
of alum (aluminium sulphate) to the water, and some kind of settling
process. Alum is not always available, and since the fluoride is
usually present only in imperceptible amounts the villagers are not
strongly motivated to treat the water or aware if it has not been
treated. When harmful or unpalatable chemicals are dissolved in the
water, it is usually preferable to look for alternative sources of water.
For instance, when ground water is salty in flat coastal areas, it is
sometimes possible to find sweet water lower down, by sinking deep
boreholes.

5.6 WATER DISTRIBUTION

Some aspects of pipeline design for rural water supplies are discussed in Section 5.1. In this section we discuss the design of rural water points.

Individual connections

Many of the potential health benefits from rural water supplies come from an increased use of water. There is therefore good reason for designing water points so as to encourage the maximum possible water use, particularly for hygiene. Ideally, water should be provided inside or near each house, as this usually leads to an increase by several times in the volume of water used, even if only a single tap is installed. The cost of a water supply with such individual connections depends on the density of village housing. It is possible to raise revenue from private subscribers, a thing very difficult to do with public water points.

Public water points

When individual connections cannot be afforded, the alternative is to provide public water points, also known as standpipes, from which the public may collect their water. In addition, showers, clothes-washing facilities, and possibly toilets may be constructed beside the water points and connected to the piped water supply. These would be provided on a communal basis, but if one shower and toilet cubicle is reserved for each family, this will help to encourage good maintenance by the users.

The best design for a public water point may depend upon traditional methods of carrying water. Where water is carried on the head, it may help if buckets can be stood on a platform at shoulder height to be filled (Figure 5.21). In that case, a lower tap should also be provided to allow clothes-washing under the tap and to permit water collection by children and old people. In designing a water point, provision should be made for the disposal of spilt water and waste water used for washing at the water point. Areas on which water will be spilt should be paved, preferably with concrete, and the waste water taken to a soakaway, such as a pit filled with stones and covered over with a layer of soil.

One serious problem is damage to water points through heavy use, or sometimes through vandalism. The most common component to break is the tap, and this should be as durable as possible. But it should not be too hard to operate; much 'vandalism' to rural water supplies is in fact the product of frustration. In any case, arrangements should be made for the regular inspection and maintenance

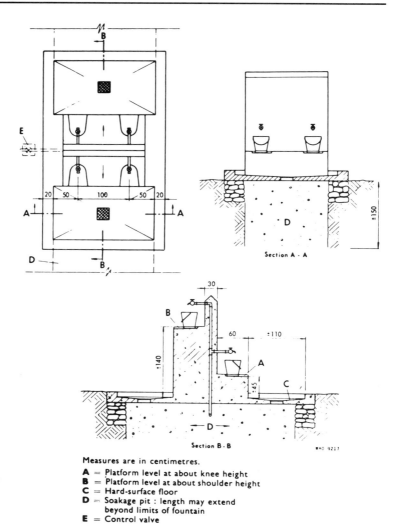

Figure 5.21 Possible design of a public water point (dimensions are in centimetres) *Source*: From Wagner and Lanoix (1959)

Measures are in centimetres.
A = Platform level at about knee height
B = Platform level at about shoulder height
C = Hard-surface floor
D = Soakage pit : length may extend beyond limits of fountain
E = Control valve

of public water points, and for villagers to report faults they cannot repair themselves.

5.7 REFERENCES AND FURTHER READING

Arlosoroff, S., Tschannerl, G., Grey, D., Journey, W., Karp, A., Langeneffer, O. and Roche, R. (1987). *Community Water Supply; the Handpump Option* (Washington DC: The World Bank).

Blankwaardt, B. (1984). *Hand Drilled Wells: a Manual on Siting, Design, Construction and Maintenance* (Amsterdam: TOOL Foundation).

Brush, R.E. (1979). *Wells Construction: Hand Dug and Hand Drilled* (Washington, DC: Peace Corps).

Cairncross, S. and Feachem, R.G. (1986). *Small Water Supplies*, Ross

Institute Bulletin No. 10 (London: London School of Hygiene and Tropical Medicine).

FAO (1977). *Self-help Wells*, Irrigation and drainage Paper No. 30 (Rome: Food and Agriculture Organization).

Glennie, C. (1983). *Village Water Supply in the Decade; Lessons from Field Experience* (Chichester: John Wiley and Sons).

Hofkes, E.H. and Visscher, J.T. (1986). *Renewable Energy Sources for Rural Water Supply*, Technical Paper Series No. 23 (The Hague: IRC International Water and Sanitation Centre).

IDRC (1981). *Rural Water Supply in China* (Ottawa: International Development Research Centre).

IRC (1979). *Public Standpost Water Supplies; A Design Manual*, Technical Paper Series No. 14 (The Hague: IRC International Water and Sanitation Centre).

IRC (1981). *Small Community Water Supplies*, Technical Paper Series No. 18 (The Hague: IRC International Water and Sanitation Centre).

Pacey, A. (1977). *Handpump Maintenance* (London: Intermediate Technology Publications).

Pickford, J. (1977). Water treatment in developing countries, in *Water, Wastes and Health in Hot Climates*, Feachem, R., McGarry, M. and Mara, D. (eds) (London: John Wiley), pp 162–191.

Rajagopalan, S. and Shiffman, M.A. (1974). *Guide to Simple Sanitary Measures for the Control of Enteric Diseases* (Geneva: World Health Organization).

Wagner, E.G. and Lanoix, J.N. (1959). *Water Supply for Rural Areas and Small Communities*, WHO Monograph Series No. 42 (Geneva: World Health Organization).

Watt, S.B. (1975). *A Manual on the Hydraulic Ram for Pumping Water* (London: Intermediate Technology Publications).

Watt, S.B. (1978). *Ferrocement Water Tanks and their Construction* (London: Intermediate Technology Publications).

Watt, S.B. and Wood, W.E. (1977). *Hand Dug Wells and their Construction* (London: Intermediate Technology Publications).

6

Urban Water Supply and Water Treatment

6.1 INTRODUCTION

Most of the technology suitable for urban water supply in developing countries is similar to that used in the developed countries, and is described in conventional textbooks on water supply such as Twort *et al.* (1985). However, by no means all of the systems used in Europe and North America are appropriate for conditions in the Third World. For example, some pumping and treatment plants could not cope with the very silty water of many tropical rivers. More importantly, some equipment is too difficult to operate, to maintain, and to repair due to difficulties of importing spare parts, shortage of trained staff, and a lack of sufficient continuity of staffing and record-keeping to ensure that the correct procedures are carried out throughout the life of the equipment. In general, therefore, the technology appropriate for urban water supply in developing countries must be chosen in such a way as to make it easily understandable by its operators, and easy to operate and repair without too much technical knowledge or need for imported materials.

Treatment is usually necessary for town water supplies. Sufficient water for a whole town is not always available from the ground, and so polluted surface sources often have to be used. The larger scale of a town water supply makes the quality of the water more important than for a small village supply. A single source of pollution in an urban supply could cause a water-borne epidemic in the whole town, so that the consequences of poor water quality are more serious. Treatment is of little use if it is only erratically applied, and yet it is a major problem to ensure continuous and reliable operation of water treatment works in many countries. Various possible methods are discussed below.

6.2 COAGULATION AND SEDIMENTATION

Because of the high silt loads of most tropical rivers, sedimentation is usually necessary as a first stage of water treatment. This involves passing the water slowly through a large tank to allow time for solid matter to settle out. It does not significantly improve the microbiological quality of the water, but makes subsequent processes more efficient.

Sedimentation is usually assisted by adding chemicals called coagulants, often alum (aluminium sulphate), to the water. This causes the small solid particles to come together in larger clusters known as 'flocs', which can settle faster through the water. The correct dose of a coagulant depends on the water being treated, and may vary from day to day. The equipment for adding the chemicals should have as few moving parts as possible, and preferably not require electricity. Some turbulence is required to mix the chemicals thoroughly with the water. This can be achieved by passing the water over a weir, through a constriction, or around baffles, and no motor-driven equipment is necessary.

Sedimentation can be accomplished in 'horizontal-flow' tanks (Figure 5.17) in which the water moves from one end to the other, or in 'upward-flow' tanks, usually circular, in which the water enters at the bottom and is taken off at the surface (Figure 6.1). Upward-flow tanks often incorporate the 'sludge blanket' or 'solids-contact' principle, by which a horizontal layer of floc develops and strains the water as it passes upwards. They are often more efficient than horizontal-flow tanks and typically have retention times in the range 1–3 hours as opposed to 4–6 hours for horizontal-flow designs. They therefore have smaller volumes, though they may not cost less.

Horizontal-flow tanks are considerably easier to build because they need not be as deep as upward-flow tanks and they have fewer internal walls. They are also easier to operate. Horizontal-flow tanks perform better with heavily silted waters than the upward-flow type, which become increasingly difficult to operate once the suspended solids exceed 1000 mg/l, as occurs in some tropical rivers. In addition, the viscosity of water is lower in warm climates, so that flocs can settle more rapidly, and large horizontal-flow tanks can be an appropriate and economical solution.

A third type combining some advantages of the other two is the spiral-flow tank, developed for treating the heavily silted water of the Nile (Figure 6.2). The water enters the tank low down on the outside wall, at an angle of 45° to the circumference, at a velocity of up to 50 mm/s. It then makes several circuits of the circular tank before flowing out over a submerged weir which occupies about a quarter of the circumference. Further details are given by Twort *et al.* (1985).

Alum-assisted sedimentation can also be used for the removal of fluoride, whose medical side-effects were mentioned in Chapter 2.

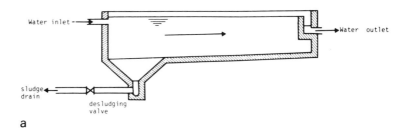

a

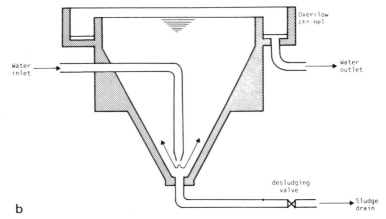

Figure 6.1 Two common types of setting tank. Many other designs exist, most of them modifications of these two basic forms (see also Figure 5.17). (a) A horizontal-flow settling tank. (b) An upward-flow clarifier

b

This is known as the Nalgonda process, and requires a much higher alum dose than is usual for conventional water treatment. The exact dose required depends on the hardness of the water and the amount of fluoride to be removed, but is typically 600 mg/l of hydrated alum, or about twenty times the typical dose for ordinary sedimentation. The prior addition of lime or sodium aluminate assists the process, and makes a lower alum dose possible. However, the method is expensive and the use of an activated alumina bed, although more complicated, would be cheaper. Both methods of fluoride removal are complicated to perform and there is no simple check that the process is being reliably operated. Surface water, although requiring treatment to remove silt and pathogens, is therefore frequently preferable to ground water containing dangerous amounts of fluoride.

The chemicals required to assist sedimentation are often applied in powder form by a 'dry feeder'. But the powder tends to 'cake' and clog the machine in humid, tropical conditions so that a solution feeder is preferable.

6.3 FILTRATION

The most commonly used method of filtration is the rapid sand filter, in which water passes downwards through a sand bed about

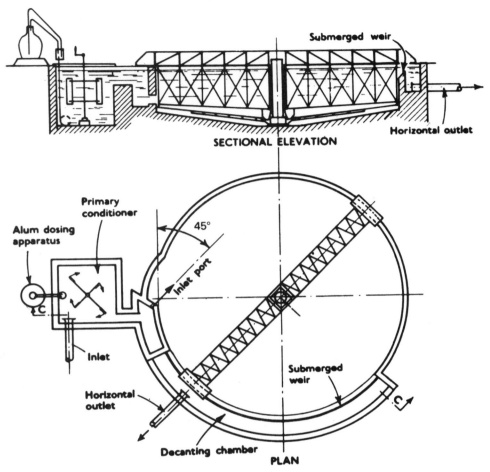

Figure 6.2   A spiral-flow settlement tank as designed by Walton and Key
Source: From Twort *et al.* (1985) Reproduced by permission of The Institution of Civil Engineers

0.45–1.0 m thick, at a rate of over 5 m/h. The water is driven through the bed either by gravity or by pressure, in which case the whole filter is contained in a steel pressure vessel. The bed requires cleaning at frequent intervals, usually at least once a day. This involves 'backwashing' by forcing water, or air followed by water, upwards through the bed for a period of time. Various modifications and simplifications in the control of the filtration rate and of backwashing have been developed (Cleasby, 1972) and are especially appropriate in developing countries.

However, because of their construction cost, their complexity, and their need for regular backwashing, rapid sand filters are inappropriate for many applications in developing countries. They are certainly not as widely suitable as the oldest and simplest method — slow sand filtration. Slow sand filters are simple to build and operate, and also improve microbiological water quality substantially, a result the other types of filter cannot reliably achieve.

Slow sand filters are so called because the water moves down through them at a rate of only about 0.2 m/h. This means that the filter beds for a large town can take up a considerable area of land, but land prices in developing countries tend to be low, so that this is not usually a severe constraint. Besides, good land on the outskirts of a rapidly growing town is a good investment for any municipal body.

Figure 6.3 shows a simple design for a slow sand filter. It consists of a large tank, in which water stands about 1 m deep over a bed of carefully graded sand. The raw water filters down through the sand to a set of underdrains, which can be made with ordinary bricks laid without mortar beneath gravel, and is collected in an outlet chamber before passing down the vertical outlet pipe. The top of the outlet pipe is fixed above the level of the sand surface to avoid negative pressures in the sand bed, and thus prevent air being entrapped.

The sand bed is 600 to 900 mm deep, but most of the filtration takes place in the top layer. At the very top of the sand bed a dense slimy layer of retained fine material develops, with an active flora and fauna. This biologically active zone, known as the 'schmutzdecke', is responsible for most of the water-quality

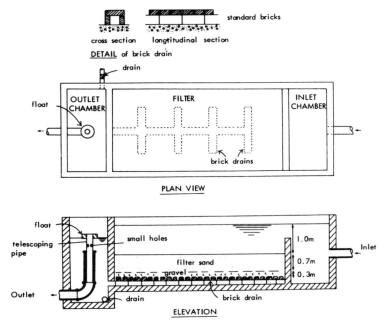

Figure 6.3   A slow sand filter

improvement provided by a slow sand filter. In particular, the schmutzdecke retains or kills the great majority of viruses, bacteria, protozoal cysts, and helminth eggs and thus makes the slow sand filter a far more efficient pathogen-removing process that the rapid sand filter.

Over a period of time, the development of the schmutzdecke increases the resistance of the filter bed to the flow of water, and it is necessary to clean it every few weeks or months. This is done by raking off 20 mm of sand from the top of the bed and discarding it. When the sand bed is only 600 mm thick, more sand is needed. The used sand can be washed in a box with water slowly piped in at the bottom, while the sand is disturbed with a spade until the water overflowing from the box becomes clean. If cleaning is required more than once a week, it means either that the sand is too fine, the flow too fast, or the water too dirty. The water may be improved before filtering by sedimentation, or by prefiltration using coarse media, such as coconut fibre or burnt rice husks.

When a filter is first used after cleaning, the water may flow through it at too high a rate. This can be prevented by fitting a floating regulator over the outlet pipe, as in Figure 6.3. The regulator starts to float when the flow through the filter becomes too great, and so raises the water level in the outlet chamber to hold the flow constant.

6.4 DISINFECTION

Slow sand filters improve the microbiological quality of water considerably, but if water completely free of pathogens is required, it is necessary to apply a chemical disinfectant. In practice, all town water supplies require disinfection. Chlorine is the disinfectant most readily available and suitable for use in most circumstances.

Chlorine demand

Chlorine is an oxidizing agent. If it is added to impure water, it will immediately oxidize the impurities and no longer be available for disinfection. It is therefore essential that the chlorine dose should be greater than that required to satisfy the immediate 'chlorine demand' of the water. This chlorine demand will vary, depending on the quality of the water. Roughly speaking, 1 mg/l of chlorine is required to satisfy 2 mg/l of BOD.

If an adequate dose is used to satisfy the chlorine demand, a 'residual' of chlorine remains which provides protection against contamination occurring during subsequent distribution of the water. Chlorine residuals are of two kinds; free and combined.

Free residuals

After the chlorine demand is satisfied the following reaction can occur:

$$Cl_2 + H_2O \rightleftharpoons \quad HCl \quad + \quad HClO$$
$$\Updownarrow \qquad\qquad \Updownarrow$$
$$H^+ + Cl^- \quad H^+ + ClO^-$$

The hypochlorous acid (HClO) and chlorite ions (ClO$^-$) are free residuals. Hypochlorous acid is the more effective disinfectant, but most of it tends to dissociate to form chlorite ions at high (alkaline) pH values. The residual may be all HClO at pH 5, about half HClO at pH 7.5 and all ClO$^-$ at pH 9. A given chlorine dose is far more effective in destroying viruses and bacteria in water if the pH is low (say pH <7). Chlorine is also considerably more effective at warm temperatures.

Combined residuals

If ammonia is present in the water, the chlorine combines with it to form chloramines, known as combined residuals. These are also disinfectants, but with less that one-twentieth the power of hypochlorous acid.

In the 1920s the practice developed of adding ammonia with chlorine deliberately to produce combined residuals. This had the principal advantage of preventing algal aftergrowths and reducing taste and odour problems. However, if used for disinfection it requires additional equipment and ammonia (an imported product in most developing countries), and a much longer 'contact time' in the water to act. Water supplies in developing countries are frequently running above their design capacity and cannot always guarantee this minimum time before the water reaches the taps. Besides, only free residuals can be effective in the removal of viruses. For these reasons, chlorine–ammonia treatment is being superseded by free-residual chlorination.

When free-residual chlorination is first introduced, the free residual may react with deposits on the inside of water mains, possibly producing disagreeable tastes and colours in the water for the first few months. However, when these deposits have been oxidized it will be possible to maintain free-residual chlorine in the water delivered throughout the distribution system.

The breakpoint

A graph of the total residual, both free and combined, against the dose of chlorine added to a water may show a characteristic curve like that in Figure 6.4. Initially, combined residuals are found. With higher doses these are destroyed by further oxidation, until a certain point is reached, known as the breakpoint, when the oxidation process is complete and free residuals begin to predominate. Thereafter, the residual increases in direct proportion to further increases in the chlorine dose.

Free-residual chlorination thus requires a chlorine dose beyond the breakpoint, and so it is sometimes called 'breakpoint chlorination'. However, many waters do not have a pronounced dip in their residual curve, and so do not have a well-defined breakpoint.

Chlorine dose

The chlorine dose must be sufficient to produce the desired free residual, after satisfying the chlorine demand. A minimum free residual of 0.3 mg/l is recommended, with a contact period of thirty minutes. Free chlorine is less effective as a disinfectant at high pH and low temperatures. A longer contact time is required in such conditions.

The chlorine demand can be estimated by adding various large doses to the water and measuring the free residuals they produce:

$$(\text{demand}) = (\text{dose}) - (\text{residual})$$

The demand and the desired residual are added arithmetically to derive the necessary dose. Even quite clean water is likely to have

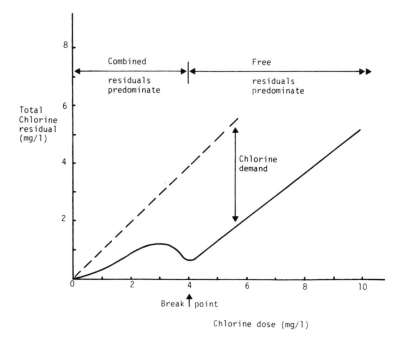

Figure 6.4   Chlorine-residual curve for a water with 0.6 mg/l ammonia
Source: After Tebutt (1992)

a chlorine demand of about 2 mg/l. When selecting chlorinating equipment, 2 mg/l of spare capacity should be included to allow for variations in chlorine demand and pH.

The chlorine demand of some waters, particularly river waters, can increase dramatically at times of heavy pollution, after rain for instance. To allow for this it may be necessary to add a safety margin to the dose. If this causes taste and odour problems, the excess chlorine can be removed, after a sufficient contact time, by adding sulphur dioxide (sulphonation):

$$Cl_2 + SO_2 + 2H_2O \rightarrow H_2SO_4 + 2HCl$$

Control of the dose

The aim of free-residual chlorination is to ensure a free residual in the water until it is supplied to the public. To check that the chlorine dose is sufficient, the residual should therefore be measured in samples taken from various points throughout the distribution system. A suitable arrangement is to check the residual in the water leaving the treatment works once every shift — two or three times a day — and to check samples from the distribution system once a week.

Strict instructions should be given that water should not be sent to the supply without an adequate residual. If not, one runs the risk of repeating the example of Kosti, a Sudanese town on the Nile, where a serious typhoid epidemic resulted when the waterworks ran out of chlorine.

Samples from the distribution system should not be taken from dead-end pipes. It is convenient to take them from institutions such as schools, fire stations and hospitals. The tap should run for several minutes before filling the sample bottle, to ensure the water is from the mains, and not the internal pipework. The absence of a free residual in only one part of the system suggests that pollution is entering the mains nearby, and should be dealt with by repairing the mains rather than increasing the dose.

The tests should be conducted at the sampling point, immediately after collecting the sample, so that the residual does not have time to change before testing.

Testing for chlorine

Simple methods are available for measuring free residuals of chlorine in water. Besides being simpler and cheaper, they are also more reliable than the sophisticated methods using photoelectric cells or continuous recorders. For this reason, the results of any automatic recorder should be regularly checked by chemical tests, such as those described below.

The two principal methods are the orthotolidine test, in which a drop of the liquid reagent is added to the water, and the more modern DPD test, using tablets. The DPD test is preferable, particularly as orthotolidine is now recognized as capable of causing cancer and its use is banned in a number of countries. Whichever test is chosen, the modified version should be used, which measures free-residual chlorine separately from the combined residual.

In both tests, the addition of a reagent produces a colouring in the water, which is checked against standard colours to measure the concentration; the stronger the colour, the higher the concentration. Orthotolidine turns the water yellow; DPD tablets turn it red. The coloured water is viewed in a test tube through a 'comparator' or 'Nesslerizer'. This is a special box in which the coloured sample tube sits next to a dummy tube containing only water, in front of which windows of coloured glass are made to pass by turning a wheel. The purpose of the dummy tube is to allow for any natural colouring in the water. However, if the water is reasonably colourless and great accuracy is not required, a cheaper colour card can be used with the DPD test instead of a comparator.

Sources of chlorine

(1) *Gas* In large-scale water treatment plants, chlorine is obtained as liquefied gas cylinders or drums. The pressure within these containers is approximately 5.5 atmospheres at 21°C. Chlorine gas is poisonous and heavier than air. If it has to be stored indoors, it should be in a building with no basement and whose doors open outwards and have a gap at the bottom. For small supplies, the complex equipment needed to apply chlorine gas and the precautions needed to handle it safely may be impractical. Other sources of chlorine are therefore used.

(2) *High-test hypochlorite* (HTH) can be obtained as a solution or in powder form. It contains up to 70% available chlorine.

(3) *Solutions* Proprietary disinfectants and bleaches may be used as chlorine sources. The disinfectants (such as Milton, Zonite, and Javel water) typically contain about 1% of available chlorine by weight. They may be used directly, without dilution, as 1% stock solutions. The bleaches (such as Chloros, Chlorox, Domestos, Voxsan, and Parazone) usually contain 3–5% of available chlorine and must be diluted to make up a 1% stock solution. Chlorine solutions are unstable in warm climates. They should be kept in brown or green bottles, well-stoppered, and stored in dark, cool places.

(4) *Powder* Bleaching powder or chlorinated lime (calcium hypochlorite) may also be used. It contains about 30% of available chlorine when fresh but the strength rapidly diminishes when the container is opened. Even when unopened, long

periods of storage can lead to reduced strength. It is best to open the container and use the entire contents immediately to make up a 1% stock solution. The inert lime will settle in a few hours leaving the active chlorine in the clear solution which can then be poured off and kept as described above.

(5) *Tablets* Various proprietary brands of chlorination tablet are on the market. They are expensive and are suitable mainly for the short-term protection of small quantities of water. Tablets of calcium hypochlorite containing 60–70% available chlorine are made for use in tablet chlorinators (see below).

The application of chlorine

Chlorine needs at least half an hour in contact with water to disinfect it. It is therefore applied before the water enters a storage tank, so that it can take effect during storage. It helps if the chlorine is applied just upstream of a weir, a water meter, a sharp bend, or some other point where turbulence of the water will help to mix the chlorine. The chlorinated water should not be exposed to sunlight after dosing, as that would remove the protective residual.

Chlorine would kill off the organisms in a slow sand filter, so that it should never be added before slow sand filtration. Besides, the chlorine demand of raw water, and its tendency to vary over time, is reduced by purification, so it is preferable to apply chlorination after other treatment processes, although chlorine is sometimes also added beforehand to control algae in the treatment works.

Chlorine gas is not applied direct to water, but used to form a strong solution in a small quantity of water which is then injected into the main stream. This ensures that the chlorine is fully dissolved. It is done by a chlorinating apparatus designed to keep the dosing rate constant and independent of the pressure in the cylinder, which varies widely with temperature.

Some chlorinators are activated by the flow through the dosed pipe, using a small by-pass from a constriction as the injector pipe. More sophisticated control devices use an automatic measurement of the downstream residual to control the input of chlorine or sulphur dioxide, but they are an unnecessary complication in a small treatment works.

For small town supplies, especially in remoter areas, chlorine is usually applied in the form of hypochlorite solution. Solution feeders, such as that in Figure 6.5, can supply a chlorine solution at a rate proportional to the flow in the main and can be operated electrically or, preferably, by the energy of the water flow provided that a minimum head is available.

Chlorine can also be applied in a solution by various drip-feed devices which can be set to add chlorine solution at an approximately constant rate (Figure 6.6). It is less easy to make them stop

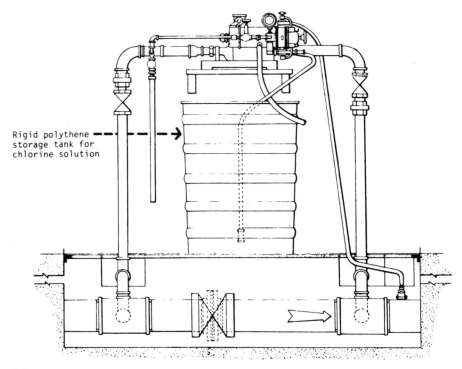

Figure 6.5   A chlorine-solution feeder. The solution is fed into the main at a rate proportional to the flow (measured by the meter on the by-pass pipe). No electricity is required and power is provided by the water itself. This model can operate with a head in the main between 7 and 90 m (Drawing: Wallace and Tiernan Ltd)

automatically when there is no flow in the dosed water pipe, and they are not normally used for dosing water at high pressure. A variety of drip-feed devices have been used other than that shown in Figure 6.6 and several are described by Assar (1971) and Pickford (1977). Those using single orifices or jets to regulate the flow are unreliable when used over extended periods, as the jets become encrusted with chloride deposit.

Several devices are available to add chlorine to water by the erosion of tablets of calcium hypochlorite (Figure 6.7). This system is convenient and easy to operate but the tablets are more costly than hypochlorite powder (from which a solution is made for application as in Figures 6.5 and 6.6) and it is sometimes difficult to guarantee the supply of tablets.

6.5 DISTRIBUTION

In many tropical cities, the water distribution system is one of the biggest headaches of the municipal engineers. As much as 60% of

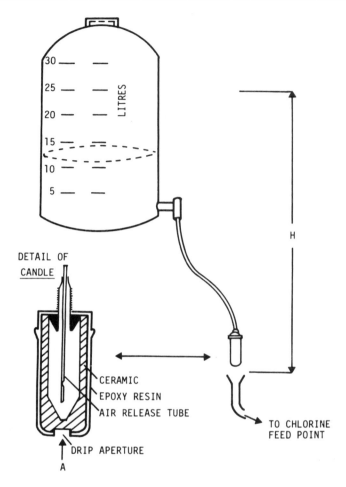

Figure 6.6 A simple drip-feed chlorinator using a ceramic filter candle. It is painted on the outside with resin and a small depression made in the tip with a 3 mm drill to penetrate the resin. The drip rate reaches a steady value within a week, and can then be adjusted by altering H or enlarging the hole at A
Source: After Barker (1967)

the water entering the system may be lost as leakage, and for much of the day the water pressure may be inadequate to reach some areas at all, let alone meet their potential fire-fighting needs. The problems of distribution systems are of four main types.

First, and this is an enormous problem in many tropical cities from Manila to Monrovia, numerous unauthorized connections are made to the water mains by private individuals. This may result from high water charges, from delays or corruption in the allocation of water connections, or from a refusal to provide connections to squatters. The resulting erosion of the water authority's revenue makes it hard to lower the charges, the leakage from badly installed unauthorized connections diverts the authority's resources (so making it difficult to shorten the delays), and attempts to police these illegal connections frequently result in increased corruption. There is no simple solution to the problem, but it is probably more

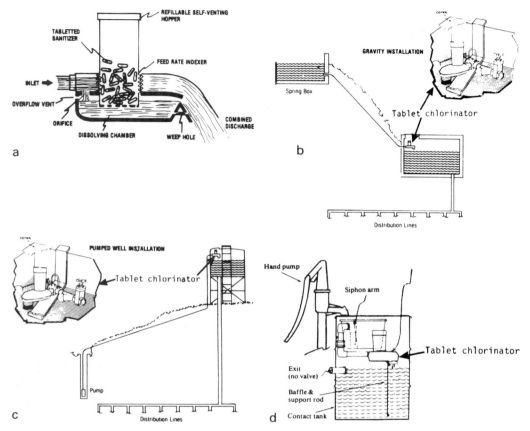

Figure 6.7   Tablet chlorinators made by World Water Resources Inc. The chlorinators use tablets of calcium hypochlorite which are eroded by the water flow as shown in drawing a. They may be installed on the storage tank of a gravity supply (drawing b) or a pumped supply (drawing c). Tablet chlorinators may also be adapted for use with a hand pump (drawing d) (Drawings: World Water Resources Inc)

constructive for the water authority to reduce its connection charges, to provide more public standpipes or to offer technical assistance to householders wishing to make their own connections.

Second, public water points are frequently damaged. Water is wasted by leaking taps, and money is wasted in repeatedly repairing them. The provision of more public water points does not necessarily increase community water use, but reduces the number of households using each. A water point serving a small group of families is more likely to be looked after and less likely to be broken by overuse or vandalism. It may even be possible to levy a water rate from them jointly. Proper institutional arrangements are necessary to ensure that it is in the interests of an identifiable individual to look after a water point (Figure 6.8). He or she may be an elected citizen, a municipal employee, or a concessionaire who pays for the right to sell water from a water point, in return for an obligation to maintain it. However, when water is sold in this way, the price paid per cubic metre is inevitably higher than the water rates paid by private consumers, since the price per bucket cannot be less than the value of the smallest coin and because an income must be guaranteed to the concessionaire. Many countries have adopted a policy of free distribution at public taps on grounds of social justice. This does not necessarily undermine the revenue which can be collected from private connections (Briscoe *et al.*, 1990). The design of public water points has been discussed in Chapter 5.

Third, the capacity of the system is often far exceeded by the water demand of the community. The rapid, but variable, growth in the urban population of most developing countries is one cause for this, but not necessarily the main one. There is little evidence that this growth in population is caused by the provision of water supply, but an improvement in the water supply can itself cause a significant increase in water demand. Some people begin to use a community water supply when the level of provision improves, having transferred from some alternative source such as a private

Figure 6.8 A water kiosk in Malawi. Water is sold to the public in an orderly and hygienic environment. A check is kept on the volume sold by the water meter, visible at the bottom left of the picture
(Photo: R Feachem)

well. Others use many times more water. While households served by standpipes use only 15–30 l/person.day, households with a single tap in the home typically use two, three or four times as much, and those with several taps use still more. Households which water their lawns may use over thirty times as much as typical households using standpipe water. However, there is substantial room for uncertainty in these figures, which could be reduced by empirical observation and measurement of existing water use, particularly at times of peak demand. When designing a distribution system, accurate population predictions are often impossible to make, but may be less important that a considered estimate of domestic water use per capita and its future rate of increase.

Fourth, high rates of leakage further overburden the distribution system. Typically, 30% of the water treated and pumped into the water supply of a developing country town, and sometimes as much as 60%, is lost in this way. It may result from incompetent pipe-laying, from exposure of mains by soil erosion, from construction of houses over mains due to poor physical planning, or from sheer age of the system. Leaky distribution mains are of particular concern with intermittent supplies and where there are low pressures, as pollution from drains and sewers may enter through the leaks when the pressure drops. Even if the overall pressure is positive, hydrodynamic effects at bends and constrictions can cause localized negative pressures sufficient to suck in pollution. Distribution mains should therefore be laid at shallower depth than any adjacent foul sewers. Repair of leaking mains is made more difficult by the lack of accurate records showing pipe routes, although the production of a set of record drawings is a useful training exercise for junior technical staff. Leakage detection methods are described in standard texts. Since the rate of leakage is roughly proportional to the pressure in the mains, a possible stopgap measure is to reduce the pressure where possible, although minimum pressures for fire-fighting must be maintained at peak times. The ultimate solution to the problem of an old and inadequate system is to replace it with one built to last. If it is completely replaced, the old distribution system may be kept for garden-watering, fire-fighting, etc.

The problem of low pressure is aggravated where there are no regulations to prevent direct pumping from water mains by consumers. This further reduces the pressure in the distribution system, and can even cause negative pressures. Pumping, to the tops of high buildings for instance, should only be permitted from storage tanks fed by gravity from the mains.

In many towns and cities, the overburdening of the system leads to its providing only an intermittent supply of water, which can have serious health consequences. The drop in pressure in a leaky distribution system when the supply is shut off can allow pollution of the supply, for instance from leaking sewers and from open drains.

Sewers, where they exist, can become blocked when the supply is off and so cause manholes to overflow with raw sewage when the supply returns. Another effect of intermittent supplies is that people collect water in larger quantities and store it for longer periods. Individuals going to public taps make fewer journeys and collect more at each journey, while those with house connections fill basins and baths with water to be used during periods when the supply is off. These practices may promote contamination of the water before it is used and therefore increase water-borne disease transmission. They may also provide new breeding sites for the mosquito *Aedes aegypti*, the major vector of epidemic yellow fever and dengue (see Chapter 15). Higher residual chlorine levels should be maintained in intermittent water supplies.

Urban water supplies are, as economists say, a 'lumpy' investment. This means that an extension to a town's water supply cannot normally be made gradually, but usually requires a lump sum expenditure. Even a small increase in the size of a pipeline requires a whole new pipeline, for instance, and the capacity of a dam reservoir cannot usually be increased without building a new dam. It is therefore economically worthwhile to plan urban water supplies to last for more than one or two decades, particularly in view of the low rates of interest on loans available for water development and the long service lifetime of modern piping.

This is particularly important where distribution systems are concerned, as their cost depends primarily on the length of pipe installed and only secondarily on its diameter. Since the health benefits of water supply are greater when water is provided in the home, diameters should if possible be chosen to pass the flows which will result from future improvements in the level of service, at least to the level of one tap in the yard of each household. When deciding on the characteristics of a distribution system, the following working rules may be helpful for estimating the relative costs of the various options.

In a typical urban water supply, the cost of the distribution system accounts for about one third of the total construction cost. A sizeable increase in flow through a pipe is permitted by a relatively small increase in its diameter. So the cost of a distribution network depends roughly on the square root of the volume of water use per capita for which it is designed. Twice as much water may be supplied through a distribution system costing only 50% more.

For a given rate of water use per capita, the length of piping in a system varies approximately with the square root of the number of taps or connections it serves. Since the additional piping required to supply more taps tends to be smaller, the total cost only increases with the fourth root. This means that, if a system with only one tap for each 500 people costs US$15 per person, it could be extended to give one tap for every 30 people (six families) at an additional cost of only $15/person. This does not include the cost of the standpipes themselves.

A progressive system of connection charges may be used in the same way as a progressive tariff system (see below) to make the richer consumers cross-subsidize the poor ones, and to cover the cost of extensions to the network.

6.6 WATER DEMAND MANAGEMENT

The important influence of per capita water demand on total community water use has already been mentioned. The most desirable form of investment in water supply improvements will not necessarily be found by treating water demand as an independent, uncontrollable factor. On the contrary, the level of service to be provided, and hence of water consumption, must be chosen in accordance with the means available to provide it. Moreover, although it is true that increased capacity is almost always necessary in the short term, it may in the longer term be possible to avoid, delay, or reduce investments in future water supply extensions while improving the level of service all round by measures to manage water demand and produce economies in water use. Such measures would bring similar benefits to waste water disposal, because less water used means less waste water to be disposed of (Chapter 8).

The demand on a water supply may be reduced without a fall in the standard of service by the following methods:

(i)   leakage reduction;
(ii)   tariff policy;
(iii)   water-saving taps and fittings;
(iv)   consumer education and information.

Leakage has been discussed above. The others are discussed in turn.

Tariff policy

Simply introducing water meters to houses with private connections can lead to substantial reductions in water demand. For example, domestic consumers in towns in the Netherlands whose water is metered use 30% less water than in those with unmetered delivery.

Water meters also permit the introduction of a 'progressive' tariff policy. This means that the largest consumers, much of whose water goes for luxury uses such as lawn-watering and car-washing, would pay more per cubic metre than the cost of supplying it. Besides helping to reduce the wasteful use of water, the surplus revenue can cross-subsidize the users of free water from standpipes as well as other poor, small consumers. A nominal charge may be levied for a minimal amount (say 10 m³/month) of water to enable the very poorest to obtain water for their basic needs, while water tariffs would increase progressively, for larger volumes. Households with only a single tap do not usually use large amounts, and could pay a nominal monthly rate, dispensing with the need for meters.

Progressive water tariffs are not a new idea; more than twenty developing countries have already put them into practice. Table 6.1 gives the pattern of water use in the capital city of an East African country and illustrates the applicability of progressive water tariffs in developing countries, where the majority of the population use relatively small quantities of water which could be subsidized by the massive consumption of the rich few. The classifications in the table refer to size and value of houses, and so permit a rough comparison between income and water demand.

The problem with meters is that they are expensive to install and replace. In most developing countries, they have to be imported; one African country found that the cost of replacement water meters was taking up half the yearly allowance of foreign exchange for the whole urban water sector. Intermittent supplies cause special problems. When the pressure comes on, the rush of air through a meter can cause it to give a false high reading. If the meter is not vandalized by an angry, overcharged consumer, it may be damaged by the impact when the water first reaches it. In Lahore, Pakistan, a World Bank study found that the average meter lasted only 5 years, and that metering would have to reduce water consumption by 80% (*sic*) for the meters to be worth the cost — assuming that those meters which worked were regularly read.

Table 6.1 Water use in an East African capital city

Consumer category	Number of connections	Average monthly consumption (m³)	Percentage of connections	Percentage of consumption
Business, institutional, and standpipes	1068	424	11	60[1]
Residential				
Type 1 (lowest income)	534	16	6	1
Type 2	5476	23	57	17
Type 3	171	43	2	1
Type 4	415	60	4	3
Type 5	1921	66	20	17
Type 6 (highest income)	61	96	1	1
Total	9646	—	100	100

[1] Standpipe consumption accounted for only 2% of the total.
Source: After Warford and Julius (1979).

An alternative form of tariff, which can also be progressive and avoids metering, though it will not constrain water demand, is one based on house values. As Table 6.1 shows, average water consumption is closely related to housing standards.

Water-saving taps and fittings

Devices are now available which can be installed in line on private connections or on individual taps and which restrict flow to a fixed amount irrespective of the pressure in the mains. These are simple to manufacture, cheap and easy to install.

Special shower nozzles and spray taps for sinks are increasingly used, and can give the same washing efficiency with less than a quarter of the flow of water from conventional fittings. They can easily be installed in existing buildings and public washing-places. A range of waste-saving taps and other control devices is described by IRC (1979b) and UNCHS (1989).

The enormous volumes of 10–20 litres which are used to flush conventional cistern-flush toilets are unnecessary for efficient operation. In the developed countries, they account for one-third to one-half of domestic water use, but can be reduced to about 3–6 l/flush, as has been shown by the widespread introduction of more efficient cisterns and flushing pans in Scandinavia (see Chapter 8). The introduction of such fittings can be assured by building regulations and bye-laws, and by government house-building policy.

Consumer education and information

Publicity campaigns to reduce water wastage and unnecessary consumption have been successful in a number of developed countries, and it is likely that, if socio-cultural preferences are taken into account, similar campaigns would prove equally effective in developing countries. Information on water supply systems could advantageously be coupled with health education aimed at schools, factories and clinics, focusing on the relationship between health, water, and excreta and sullage disposal.

6.7 REFERENCES AND FURTHER READING

Assar, M. (1971). *Guide to Sanitation in Natural Disasters* (Geneva: World Health Organization)

Barker, M.W. (1967). A drip-feed system for chlorination of small water supplies. *Journal of the Society of Water Treatment and Examination*, **16**, 34.

Briscoe, J., de Castro, P.F., Griffin, C., North, J. and Olsen, O. (1990). Toward equitable and sustainable rural water supplies; a contingent valuation study in Brazil. *World Bank Economic Review*, **4**(2), 115–134.

Cleasby, J.L. (1972). Filtration, in *Physicochemical Processes for Water Quality Control*, Weber, W.J. (ed.) (New York: John Wiley).

Cox, C.R. (1964). *Operation and Control of Water Treatment Processes*, WHO Monograph Series No. 49 (Geneva: World Health Organization).

IRC (1978). *Slow Sand Filtration for Community Water Supply in Developing Countries: A Design and Construction Manual*, Technical Paper Series No. 11 (The Hague: IRC International Water and Sanitation Centre).

IRC (1979a). *Public Standpost Water Supplies*, Technical Paper Series No. 13 (The Hague: IRC International Water and Sanitation Centre).

IRC (1979b). *Public Standpost Water Supplies: A Design Manual*, Technical Paper Series No. 14 (The Hague: IRC International Water and Sanitation Centre).

Lauria, D.T., Kolsky, P.J. and Middleton, R.N. (1979). Design of water systems for developing countries, *Progress in Water Technology*, **11**, 151–157.

Pickford, J. (1977). Water treatment in developing countries, in *Water, Wastes and Health in Hot Climates*, Feachem, R., McGarry, M. and Mara, D. (eds) (London: John Wiley), pp. 162–191.

Schulz, C.R. and Okun, D.A. (1984). *Surface Water Treatment for Communities in Developing Countries*. (New York: John Wiley).

Smethurst, G. (1979). *Basic Water Treatment for Application World-Wide* (London: Thomas Telford).

Tebbutt, T.H.Y. (1992). *Principles of Water Quality Control*, 4th edition (Oxford: Pergamon).

Twort, A.C., Hoather, R.C. and Law, F.M. (1985). *Water Supply*, 3rd edition (London: Edward Arnold).

UNCHS (Habitat) (1989). *The Conservation of Drinking-water Supplies; Techniques for Low-income Settlements*. (Nairobi: UN Centre for Human Settlements).

Wagner, E.G. and Lanoix, J.N. (1959). *Water Supply for Rural Areas and Small Communities*, WHO Monograph Series No. 42 (Geneva: World Health Organization).

Warford, J.J. and Julius, D.S. (1979). Water rate policy: lessons from less developed countries, *Journal of the American Water Works Association*, **77**, 199–203.

Part III

Excreta and Refuse: Treatment, Disposal and Re-use

7

Excreta Disposal in Developing Countries

7.1 THE HAVES AND THE HAVE NOTS

The World Health Organization's figures for 1988 showed that only 67% of the combined urban population of the developing countries had adequate facilities for excreta disposal. Only a minority of these were served by piped sewerage systems. In the rural areas, only 19% had adequate excreta disposal facilities (Figure 7.1). Figure 7.2 shows the levels of service in the rural areas of various countries. These figures are based on very modest definitions of adequacy. Many latrines do not meet minimal public health requirements, are not accessible to children, or are liable to pollute nearby wells.

7.2 MARKETING OF LOW-COST SANITATION

To those for whom water supply and excreta disposal both imply pipes laid beneath the street, there are obvious advantages in combining them in a single programme. However, in the context of low-income communities in developing countries their technology and their manner of implementation is fundamentally different. In such a setting, a water supply means a tap in the street or a pump in the village square, clearly in the public domain. Sanitation, on the other hand, usually means a toilet with an on-site disposal system, a part of the owner's house, built on his land, at his expense and frequently with his (or her) own hands. Its use requires a change in some very intimate habits, in the privacy of the home, by all members of the family. Whereas water supplies are almost universally popular, sanitation facilities are unlikely to be used, still less maintained, unless people want them.

Sanitation therefore has to be marketed, and this requires a very different approach from conventional civil engineering.

(1) In general, health improvement does not motivate many people

Figure 7.1 In the developing countries about half the urban population and nearly all the rural population lack adequate disposal facilities. In rural areas (a) people typically defaecate near their houses or in the fields. Children defaecate in the yard. In urban areas (b) wealthy people have flush toilets while the urban poor may have no toilets or latrines at all
(Photos: A Almary, WHO)

to buy latrines, because the connection between latrine usage and health is not clearly perceived. The desire for privacy, convenience or social status is usually more effective in generating demand.

(2) The cost is not a function of the design criteria; rather, the design criteria should depend on the price which purchasers are willing to pay (Figure 7.3); while some programmes have offered latrines at heavily subsidized prices, most of these have reached only a tiny percentage of the target population.

(3) A modification to an existing practice or type of latrine is likely to be much easier to implement than a completely new package of technology. Before marketing a new product, it is essential to study what people already do, and ask them what they think they need.

(4) The acceptability of the product (the sanitation technology) must be checked at every stage in its development by consulting likely purchasers, and by offering prototypes to some of them. It is a good idea to offer a range of models.

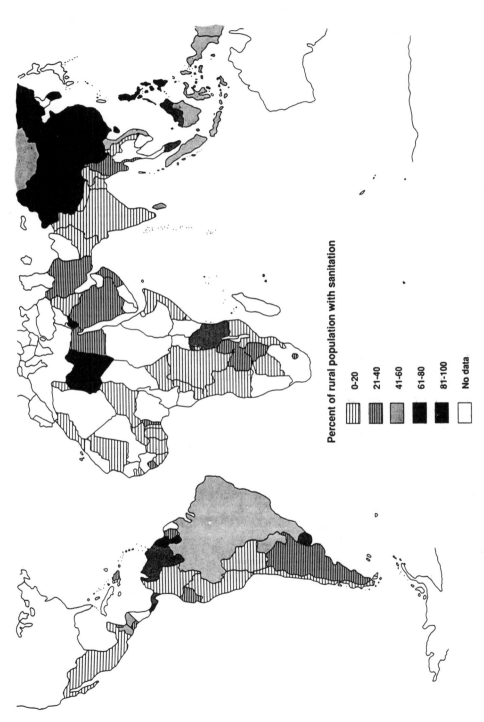

Figure 7.2   Reported percentages of the rural population of various countries having adequate excreta disposal facilities in 1988. These figures are optimistic estimates, and some countries have lower coverage than shown.

(5) The marketing operation requires constant monitoring of the consumer's response. Sanitation promotion, including this monitoring, is best accomplished through a cadre of staff in direct contact with the consumers in the field.

(6) The rate of installation depends on demand, and not on any preconceived project plan. Demand may take several years to build up, as many people will wait until their neighbours have installed a latrine and found it to perform satisfactorily, before they buy one for themselves. The most successful and sustainable sanitation programmes have been led by consumer demand.

(7) There must be someone to provide 'after-sales service' if the technology is not to become discredited. Without good maintenance, any type of latrine soon becomes fouled and offensive, and may then become a health hazard in itself.

7.3 URBAN SANITATION

A major challenge facing those concerned with environmental health in developing countries is that of excreta and refuse disposal systems appropriate to high-density, low-income communities. Such communities are found in increasing numbers around all major towns and cities in developing countries. High-density communities without adequate sanitation range from the totally unplanned squatter settlements and slums (which several governments have tried with little success to prohibit or destroy) to planned high-density housing areas where adequate sanitation has not been provided, largely due to the absence of an acceptable system for which the community could afford to pay.

Figure 7.3 Simple pit latrines, such as this one in Kenya, are an acceptable rural excreta disposal system if they are used by all members of the family and kept clean. Rural people may be encouraged to install facilities of this type by government programmes of hygiene education and the distribution of mass-produced latrine slabs at subsidized cost or on credit
(Photo: R Feachem)

The sanitation system which is by far the most convenient to the user is the conventional water-borne sewerage system found in most European communities. However, there are several reasons why a water-borne sanitation system is inappropriate for most high-density communities in developing countries.

Cost   The water-borne system is the most expensive of all sanitation systems and has a very high capital construction cost. The cost of laying the sewers alone may be as high as US $1300/person. Assuming that a city authority cannot obtain donor aid to cover the capital costs, then the money must be borrowed and repaid later. Either the community must cover the repayment costs by sewerage charges or additional taxes, or the city must subsidize the sanitation sector at an opportunity cost to other sectors of possible public spending. Experience has shown that most high-density, low-income communities are unable or unwilling to cover the real capital and running costs of water-borne sewerage and that city and town authorities are reluctant to subsidize urban sanitation for the poor.

Water use   Water-borne systems use large volumes of drinking water merely to transport wastes along pipes — water which has to be expensively treated before being released back into the hydrological cycle. This extravagant use of water may be justified in a country with ample water resources and a well-established distribution system. It is not justified in many developing countries, where water is scarce and expensive and where distribution systems are very limited and frequently overloaded. Moreover, many developing countries are arid for at least a few months of each year. During these periods even less water is available and there may be no flowing rivers into which to discharge effluent.

Water-borne systems can be installed only in communities with individual-house water connections; the majority of low-income urban dwellers do not have this facility. Many of those with connections have only a single tap in their house and many have only an intermittent supply of water. Sewers can rapidly become blocked during the periods when the water is shut off. It is not just a question of the adequacy of the distribution system, but also the quantity of the water available. Communities with water-borne sewage normally require more than 75 l/person.day, compared with less than 20 l/person.day currently used in many squatter settlements.

Construction   Water-borne sewerage is a complex technology requiring careful and skilled construction if it is to operate smoothly. The skills necessary to design and install such a system may be in very short supply in a developing country, thus forcing the employment of expatriate companies, with consequent loss of foreign exchange.

Sewer-laying   By and large, sewers must be laid in straight lines. To dig trenches in straight lines through squatter settlements necessitates the demolition of a substantial number of houses, which is often be politically and socially unacceptable.

Sewers must be laid to a constant falling gradient. On the flat alluvial plains on which many tropical cities are built, this means that numerous pumping stations are required if the pipes are not to be laid excessively (and expensively) deep below ground. These add to the cost and the maintenance problems of the system, especially where the electricity supply is unreliable.

Blockage   Conventional water-borne systems are prone to blockage if large objects are fed into them, or if inadequate water is available for flushing. Communities unused to water-borne sewerage will often try to use the system to remove a variety of household wastes, some of which will block the sewers. Materials used traditionally in certain areas for anal cleansing, such as corn cobs and stones, may also obstruct sewers.

By far the biggest obstacle to the adoption of water-borne sewerage for the urban poor is cost. For example, the World Bank has estimated that the construction and operation costs of a sewerage system per household per year amount to roughly ten times the cost of an improved pit latrine or pour-flush toilet.

There is no single most hygienic and appropriate alternative to conventional water-borne sewerage for the urban poor. Various systems are required to suit the diverse environments into which they are to be introduced. Several technologies have been proposed, of which the most promising are discussed in the following chapter. Some of these were invented many years ago; others are much more recent.

7.4 REFERENCES AND FURTHER READING

Cairncross, S. (1988). *Small Scale Sanitation*, Ross Institute Bulletin No. 8 (London: London School of Hygiene & Tropical Medicine).

Cairncross, S. (1992). *Sanitation and Water Supply: Practical Lessons from the Decade*, Water and Sanitation Discussion Paper No. 9 (Washington DC: World Bank).

Feachem, R.G., Bradley, D.J., Garelick, H. and Mara, D.D. (1983). *Sanitation and Disease: Health Aspects of Excreta and Wastewater Management* (London: John Wiley).

Kalbermatten, J.K., Julius, D.J. and Gunnerson, C.G. (1982). *Appropriate Sanitation Alternatives: a Technical and Economic Appraisal* (Baltimore: Johns Hopkins University Press).

Rybczynski, W., Polprasert, C. and McGarry, M. (1978). *Low-cost Technology Options for Sanitation: a State-of-the-Art Review and Annotated Bibliography* (Ottawa: International Development Research Centre).

Wagner, E.G. and Lanoix, J.P. (1958). *Excreta Disposal for Rural Areas and Small Communities* (Geneva: World Health Organization).

World Bank (1980). *Water Supply and Waste Disposal*. Poverty and Basic Needs Series (Washington DC: World Bank).

8

Types of Excreta Disposal System

8.1 INTRODUCTION

As pointed out in Chapter 7, the use of water-borne sewerage is often impracticable in both urban and rural contexts in developing countries. The design and construction of water-borne sewerage is well described in many standard texts on waste-water engineering (for instance, Okun and Ponghis, 1975) and so is not dealt with here. In this chapter we consider the options for sanitation improvements at low cost.

8.2 PIT LATRINES

The simplest and cheapest improvement to a pit latrine is to provide it with a prefabricated floor, in the form of a squatting slab (Figure 8.1) or with a seat. This has the following advantages:

- the latrine will be structurally safer and (no less important), it will *feel* safer;
- it will be easier to clean;
- using the footrests, it will be easier for users to position themselves over the drop hole, so as not to foul it;
- a hole with the dimensions shown is too small for a child to fall into it, and is therefore safer and less frightening;
- the cement floor will prevent hookworm transmission;
- it also permits a small measure of fly control, through the use of a tight-fitting lid.

The need for steel reinforcement can be reduced or even avoided by making the slab slightly domed or conical in shape (Figure 8.2). Alternatively, a latrine with a strong floor made with local materials such as wood and earth can be improved by placing a small slab, 60 cm square, over the centre. Since this 'finishing slab' is not a

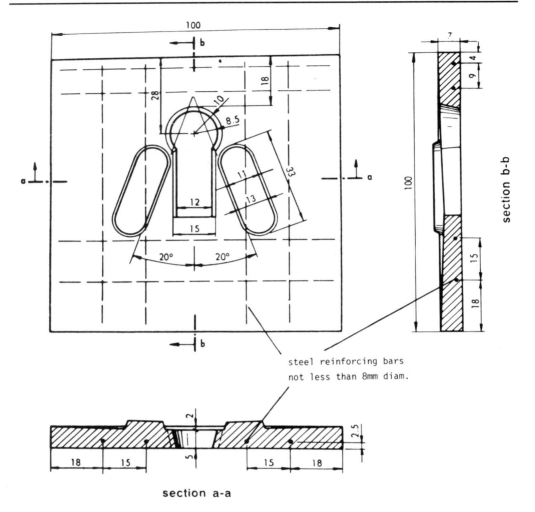

section b-b

steel reinforcing bars
not less than 8mm diam.

section a-a

Figure 8.1   Square, reinforced, concrete squatting slab for a pit latrine (dimensions are in centimetres)
Source: From Wagner and Lanoix (1958)

structural bridge over the pit, it needs no reinforcement and can be
only 40 mm thick.

The pit can be greatly strengthened against collapse by building
a lining or a ring beam around the upper part. A ring beam can be
cast in concrete *in situ*, possibly in a carefully excavated 0.2 × 0.2 m
trench before digging the pit. A lining is particularly important in
loose soils or where the pit will be full of water, but should not
prevent the seepage of fluids into the ground.

The pit should be as large as possible. The pit volume should
be at least 0.06 m³/person for every year of anticipated life, not
including the top 0.5 m, as the pit must be emptied or filled with

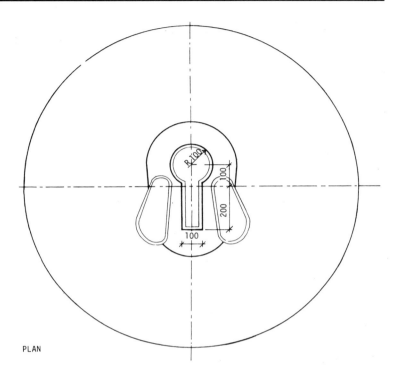

PLAN

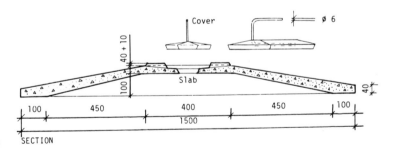

SECTION

Figure 8.2 A round, conical, unreinforced concrete squatting slab developed in Mozambique (dimensions are in millimetres) (Drawings: B Brandberg)

earth before it is completely full. A further 50% should be added where bulky materials, such as stones, maize cobs, or cement bags are used for anal cleaning. When the excreta in the pit will be under water, the volume may be reduced to 0.04 m³/person.year.

Where ground conditions allow and where there is no danger of contaminating water supplies (see Section 8.4), the need for a structural cover may be avoided by drilling a hand-augered borehole about 6 m deep and 400 mm in diameter instead of digging a pit. Although the borehole latrine has a smaller volume and therefore fills up faster than a pit, it is quicker to install in large numbers and requires only a small and relatively portable slab. It is particularly appropriate following disasters where large numbers of latrines must be rapidly installed.

In some countries, a seat is more acceptable than a squatting slab in spite of its higher cost. It also has the advantage that it is less likely to be fouled by excreta which miss the hole. Moreover, when a toilet with a squatting slab is cleaned, grit and other debris on the floor will be swept down the hole, tending to fill up the pit prematurely.

Flies often lay their eggs in faeces, and poorly built latrines can lead to an increase in the population of flies carrying faecal pathogens. Pouring insecticides into the pit to kill flies is not recommended. Although it will kill flies in the short term, it may permit a still greater resurgence later, by killing the flies' competitors and predators.

Insecticides have been used successfully in this way to control mosquitoes in flooded latrines. But their use in a continuous programme is not recommended. It is likely that the regular spraying of latrines in Dar-es-Salaam, Tanzania, during the 1970s was a major cause of the development of resistance to organophosphate insecticides in the local *Culex pipiens* mosquitoes. There is also a danger of the insecticides polluting nearby water sources.

A simple fly-control measure is to drape pieces of sacking soaked in used motor oil around the edges of the pit, hanging down by at least 300 mm and extending out horizontally under the floor by at least 500 mm. This interrupts the emergence of flies hatched from the excreta, which spend a period of their development buried in the pit wall. More sophisticated fly- and mosquito-control methods, based on ventilation pipes with a mosquito screen, are discussed below.

8.3 VIP LATRINES

The two principal disadvantages of the conventional pit latrine — namely that it smells and produces hundreds of flies (or mosquitoes) a day — are reduced in the types known as ventilated improved pit (VIP) latrines. The single-pit version is shown in Figure 8.3. Due to the action of wind passing over the top of the vent pipe, the air inside rises and escapes to the atmosphere, so creating a downdraught of air through the squatting plate or seat. This circulation of air effectively removes the odours emanating from the faecal material in the pit.

The vent pipe also has an important role to play in fly control. Female flies, searching for an egg-laying site, are attracted by the odours from the vent pipe but are prevented from flying down the pipe by the fly screen at its top (Figure 8.3). Nonetheless, some flies may enter via the drop hole and lay their eggs. When new adult flies emerge they instinctively fly towards the light; however, if the latrine is suitably dark inside the only light they can see is that at the top of the vent pipe. If the vent pipe is provided with a fly screen at its top, the new flies will not be able to escape and they will eventually fall down and die in the pit. In controlled experiments in Zimbabwe (Morgan, 1977) 13 953 flies were caught during a 78-day period from an unvented pit latrine, but only 146 were caught from a vented (but otherwise identical) pit latrine. The fly screen should have a 1 mm mesh, and should preferably be of stainless steel or glass fibre so as to resist the corrosive gases emerging from the pit. Once a year, a bucket of water should be poured down the pipe to clear it of cobwebs, and the fly screen checked, and replaced if it is not intact.

This simple measure is not so effective, however, against the *Culex pipiens* mosquitoes which breed in flooded pit latrines and can spread filariasis. The mosquitoes are less attracted by the light because they emerge at dusk and seem more able to find alternative escape routes via the squatting slab or any small opening (Curtis and Feachem, 1981). An additional device, effective against flies and mosquitoes, is a fly trap (Figure 8.4) placed over the drop hole instead of a cover. It may also be used on unventilated latrines. Provision must be made for removing the corpsès, or a small lizard might be kept inside the trap to scavenge on the insects. In practice, a more effective treatment to prevent mosquito breeding in a pit latrine is the polystyrene bead method described in Section 15.2.

Although the single-pit VIP latrine can be designed with a long life (one or more years) and to permit it to be emptied so that it can be a permanent structure, it is often more convenient and possibly less expensive to design a twin-pit VIP latrine of the type shown in Figure 8.5. In this version one pit is used for a given period (at least twelve months) until it is full, when the second pit is put into use; when that is full, the first is emptied and used again. Thus the excreta are never handled until they are at least twelve months old, when only a few *Ascaris* eggs at most will still be alive. No other materials are added to the pits, which both act as normal leaching pits. The limited period between emptying permits the use of smaller pits than for a conventional pit latrine, though at least two years' capacity in each pit is advisable. Unvented twin-pit latrines are traditional in some parts of the world, for example, in the state of Santa Catarina in Brazil; the addition of vent pipes, one for each pit, is relatively inexpensive and reduces fly and odour nuisance.

The use of VIP latrines in urban areas presupposes of course, the existence of a pit-emptying programme. The emptying of a pit latrine by mechanical means is not easy. If a vacuum tanker is to be used, water must usually be added first and the pit contents mixed into a slurry. In this case, the pit should have a lining. Householders may empty their own pits by hand if they are able and keen to use the humus-like material as fertilizer on their plots and if they do not consider this operation unacceptable. Pit emptying may alternatively be a municipal function, which in practice may well have adminis-

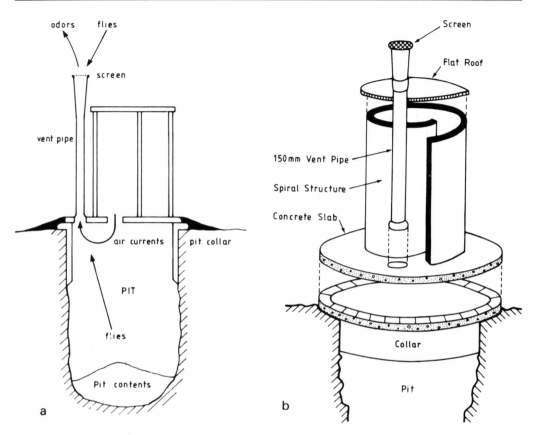

Figure 8.3   These pictures show VIP latrines as developed in Zimbabwe. (a) Schematic diagram of a VIP latrine (b) Exploded schematic diagram of a VIP latrine with a spiral, ferrocement super-structure (c) A VIP latrine with a spiral, ferrocement superstructure and an asbestos cement vent pipe (d) Exploded schematic diagram of a VIP latrine with a spiral, mud and wattle superstructure (e) A VIP latrine with a spiral, mud and wattle superstructure and a plastered reed vent pipe
Source: From Morgan and Mara (1982)

trative difficulties (Chapter 9), or it may be carried out by the private sector, for example by local farmers or by a private concern which sells the material to farmers or otherwise disposes of it. In any event, the arrangements for removing and disposing of the pit contents must be planned before VIP latrines are built on a wide scale.

8.4 DIFFICULTIES WITH PIT LATRINES

The principal areas in which difficulties are encountered in the construction of pit latrines (and of other forms of sanitation in which excreta are disposed of 'on-site' in the ground) are those with rock or sandy soil, those with high water tables, and those where there is a danger of contaminating nearby water supplies.

c

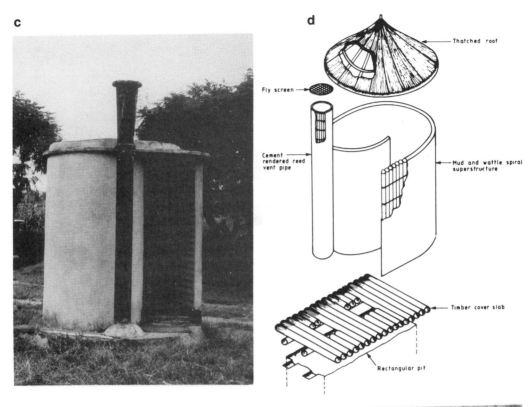

d

Thatched roof

Fly screen

Cement
rendered reed
vent pipe

Mud and wattle spiral
superstructure

Timber cover slab

Rectangular pit

e

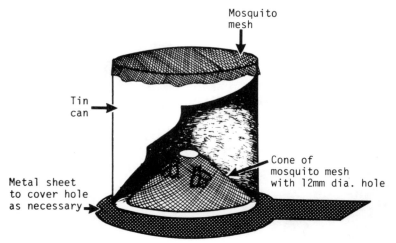

Mosquito mesh

Tin can

Metal sheet to cover hole as necessary

Cone of mosquito mesh with 12mm dia. hole

Figure 8.4 Cutaway view of simple fly trap made from an old paint tin. It is placed over the drop hole of ventilated or unventilated latrines to catch young flies and mosquitoes that try to leave the pit

Figure 8.5 Construction of a twin-pit VIP latrine. In practice, it is preferable to have two vent pipes, one for each pit *Source*: After Carroll (1980). Reproduced by permission of the Controller, HMSO.

Rocky ground

Pit latrine construction becomes both difficult and expensive in rocky ground. There is no easy answer to this except that householders wishing to build pits in rocky areas may appreciate assistance from the local public works department with mechanical diggers. The temptation in rocky ground is to build very small pits, which quickly become filled so that more pits are required. This temptation should be resisted; the principle of building pits as large as possible should apply even more strongly in rocky areas.

Sand

Pits dug in loose and unconsolidated soils (e.g. sand or fine-grained alluvium) are liable to collapse. They will therefore need to be supported as in Figure 8.6. The lining must not prevent the seepage of faecal liquors out of the pit onto the surrounding soil.

High water table

Where the level of water in the ground is high, the construction of pits becomes very difficult; they tend to collapse in the wet season, and there is a danger of mosquitoes breeding in pits with high water levels (see Chapter 15). In such circumstances a built-up pit is appropriate, as shown in Figure 8.6. The raised part of the plinth should be plastered on the inside and soil from the pit placed against the outside, to prevent leakage of fluids above ground and erosion by storm water.

Water contamination

The placing of wastes in pits always presents the danger of polluting water sources — particularly wells located nearby, and also water mains with low or intermittent pressure (Chapter 6). The danger of pollution is increased if the pit is dug down to the water table or to fissured or weathered rock. As a general rule, pit latrines should not be built within 10 m of a well or other drinking water source and should not be located uphill from it. Greater distances are appropriate when large volumes of water are abstracted from the well by a motorized pump (Lewis *et al.*, 1982).

Bacteria will not penetrate more than 1–2 m in most unsaturated soils, but they have been known to travel over 100 m in gravel below the water table and in rock fissures. In general, the bacterial contamination may spread as far as the distance travelled by the ground water itself in ten days; chemical pollution, for example by nitrates, may travel further.

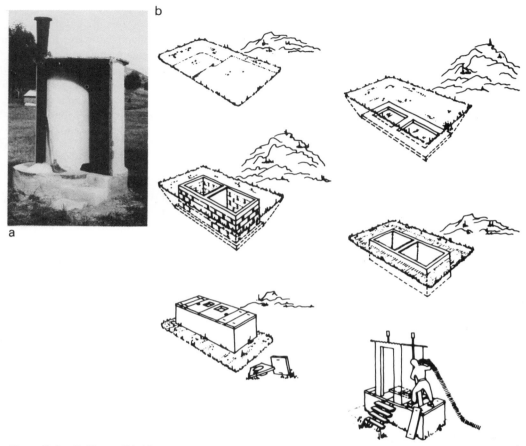

Figure 8.6   Built-up pit latrines. (a) A built-up VIP latrine in Zimbabwe in an area with high ground water table
Source: From Morgan and Mara (1982)
(b) Construction of a built-up, twin-pit, VIP latrine
Source: From Carroll (1980). Reproduced by permission of the controller, HMSO

Where it is necessary to avoid any risk of faecal contamination of ground water, there should be at least 2 m of soil depth between the pit floor and the water table or rock surface. In areas where pit latrines are unavoidably located near wells and boreholes, the water quality should be regularly checked for the presence of faecal bacteria and nitrates (see Chapter 3).

In areas of high population density and low rainfall, the rainfall may not be sufficient to flush from the groundwater the nitrates accumulated from latrines or leaching pits, and nitrates may build up to dangerous levels. A simple calculation to evaluate the potential importance of this problem may be illustrated by the example of a village in Botswana, with a population density of 63 people/hectare. A person excretes about 5 kg of nitrogen/year. It is reasonable to assume that only 20% of the total nitrogen in the faecal matter deposited in the pits is leached to the water table. The total input per annum in this village can then be estimated:

$$63 \times 5 \times \frac{20}{100} = 63 \text{ kg nitrogen/ha}$$

This amount of nitrogen would require 630 mm of infiltration to dilute the recharging water to the WHO maximum acceptable concentration of 10 mg/l of nitrogen in nitrates. Since the average annual rainfall in that region is approximately 500 mm/year, of which probably only 50 mm infiltrate into the ground, it is no wonder that there is a severe local nitrate pollution problem in that part of the world.

The risk of groundwater pollution is not necessarily a reason to reject on-site excreta disposal options. It is usually cheaper to provide an off-site water supply than off-site sanitation.

8.5 POUR-FLUSH TOILETS

A further improvement to the pit latrine can be obtained with a water seal, which is a U-pipe filled with water, below the seat or squatting pan (Figure 8.7), and which completely prevents the passage of flies and odours. By careful design of the pan, with a water seal only 15–25 mm deep, the very high water demand of the conventional cistern-flushed system can be avoided, and the toilet flushed by hand.

The smaller quantities of water used in pour-flushed toilets are not sufficient to operate a conventional system of sewerage, but are enough to carry the excreta to a soakage pit up to 8 m away (Figure 8.8), which will be easier to empty than a pit directly beneath the seal.

Pour-flush toilets are very common in the Indian subcontinent and the Far East. They have three main advantages: low water requirements (1–3 l/flush as opposed to 10–20 l/flush for most cistern-flush toilets); complete odour and fly elimination by the shallow water seal; and they can be located, if desired, inside the house, and not necessarily only on the ground floor. They are particularly suitable wherever water is used for anal cleansing. Since flushing is done manually, they do not require a multiple-tap in-house level of water supply; they are thus best used in conjunction with a yard-tap level of water supply, although they can be used in conjunction with public standpipes if the standpipe density is such that the users can and will carry enough water home for their operation.

As in the case of VIP latrines, probably the better long-term solution is for pour-flush toilets to have two pits which are used alternately (Figure 8.8), although this depends on the ease with which the pits can be desludged, whether desludging is to be done manually

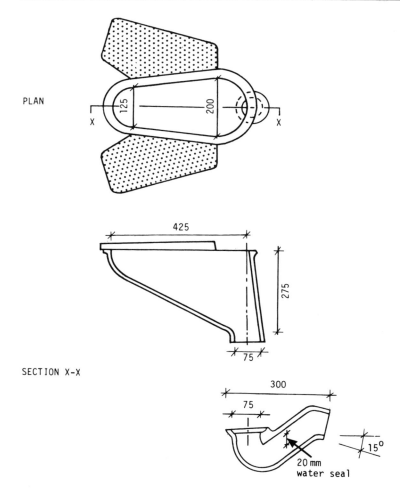

PLAN

SECTION X-X

Figure 8.7 Pour-flush squatting pan and trap developed by the Indian stands Institution (dimensions are in millimetres) (Drawing: A K Roy)

or mechanically and whether in high-density areas there is sufficient room for twin pits. If desludging is to be done by hand, then to protect the health of the person carrying out this operation and to avoid the need for sludge treatment, twin pits each with a life of at least twelve months are preferable.

If the soil conditions are not suitable for disposal to a soakage pit, a pour-flush toilet is still feasible, but in this case it should discharge into a small two-compartment septic tank (see below); to reduce costs the septic tank may be shared by two or more adjacent houses. The first compartment receives only the pour-flush waste water: after settlement this passes into the second compartment which also directly receives all the sullage. This strategy (which has yet to be tried in practice) would ensure that the septic tank effluent contains fewer faecal solids. The effluent may then be discharged into a small-bore sewer (see below) or a covered stormwater drain.

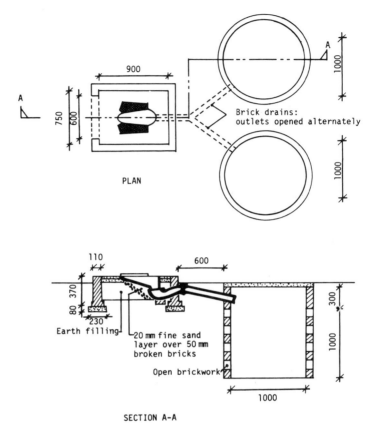

Figure 8.8 Twin-pit pour-flush toilet developed by the World Bank/UNDP Technology Advisory Group for urban sanitation projects in India (dimensions are in millimetres) (Drawing: A K Roy)

8.6 VAULT TOILETS AND CARTAGE

The vault toilet, popular in Japan and other countries in the Far East, is essentially a pour-flush toilet discharging into a watertight vault which stores the toilet waste water for two to four weeks (Figure 8.9). It is then removed by a vacuum tanker and taken away for treatment.

In general, all cartage systems which require manual removal of nightsoil from one vessel to another are extremely unhygienic and offensive, and some sort of pumping system is to be preferred. Cartage can then be a hygienic form of nightsoil removal, although it has high operating costs and demands a high level of municipal organization. It is a very flexible form of sanitation compared with conventional sewerage: changes in land use patterns (for example, from high-density low-income residential to industrial usage) are easily accommodated by merely rerouting the tankers.

Access to individual houses and street congestion may create difficulties in areas of high population density, but the vacuum tanker does not have to be a large, expensive vehicle; animal-drawn carts

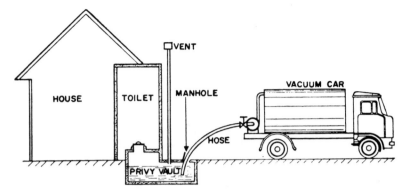

Figure 8.9 Typical arrangement for household nightsoil collection with vault and vacuum truck

with small tanks and manually operated vacuum pumps could be a perfectly feasible alternative. In general, operating costs should be significantly reduced by the use of appropriately designed systems, even though this may require considerably more ingenuity on the part of the design engineer.

8.7 SEPTIC TANKS

A septic tank is a watertight settling tank to which wastes are carried by water flushing down a short sewer. A septic tank does not dispose of wastes; it only helps to separate and digest the solid matter. The liquid effluent flowing out of the tank remains to be disposed of, normally by a soakage pit or drainfield, and the sludge accumulating in the tank must be periodically removed.

Septic tanks are generally of the single- or, better, the double-compartment variety; in conventional design practice they receive the effluent both from cistern-flush toilets and all the household sullage. In double-compartment tanks the first compartment, which receives both types of waste water, has twice the volume of the second (Figure 8.10). For a single household, the total volume of the tank should be a least three times the average volume of water used daily. More modest design criteria for larger septic tanks are given by Mara and Sinnatamby (1986). The conventionally designed septic tank works well in low-density areas (less than about 100 persons/ha) where the soil conditions are suitable, but at higher densities there is insufficient space for adequate drainfields. In many tropical cities it is common to see septic tank effluents overflowing from overloaded soakaways and discharging to stormwater drains, often open and frequently blocked.

By modifying the design, it should be possible to use septic tanks at higher densities, provided of course the soil is suitable for on-site disposal. The design modification is to give the septic

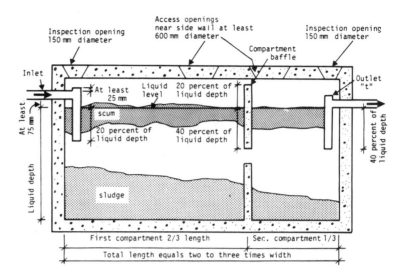

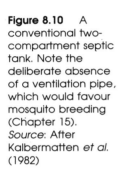

Figure 8.10 A conventional two-compartment septic tank. Note the deliberate absence of a ventilation pipe, which would favour mosquito breeding (Chapter 15).
Source: After Kalbermatten *et al.* (1982)

tank three compartments (Figure 8.11); the first receives only the cistern-flush toilet waste water which after settlement passes to the second compartment for further settlement and thence into a third compartment which also directly receives all the household sullage.

The advantage of this strategy is the same as for the sewered pour-flush toilet — the effluent contains fewer faecal solids; additional settlement for the toilet waste water is provided in the second compartment as the hydraulic disturbance in the first compartment caused by the discharge of the cistern-flush toilet is much greater

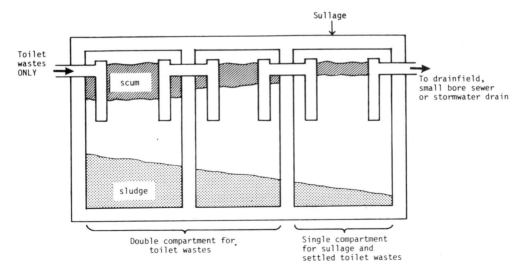

Figure 8.11   Three-compartment septic tank
Source: After Kalbermatten *et al.* (1982)

than with a pour-flush toilet and the carry-over of solids into the next compartment is correspondingly greater. The net result of having three compartments and initially separating the toilet waste water and the sullage is that the effluent can be expected to have a long-term infiltration rate some two to three times greater than the effluent from a conventionally-designed septic tank, so that the drainfield can be two to three times smaller.

Thus the modified septic tank could be used at higher densities, at least 200 persons/ha and possibly 300 persons/ha, obviating the need for conventional sewerage in areas with these densities. If the soil conditions are not suitable for on-site disposal by a drainfield, then a small-bore sewerage system to receive the septic tank effluent should be considered; in any given situation it is simple to determine whether such a system has a lower economic cost than conventional sewerage. The main point is, as always in sanitation programme planning, that all feasible alternatives should be examined and the most economically, socially and environmentally appropriate one adopted.

8.8 SMALL-BORE SEWERS

Small-bore sewers, also known as 'effluent drains', have been in use for many years in several communities in the USA and Australia. They carry only the effluent from septic tanks (Figure 8.12). Since the large solids have been removed from the waste by sedimentation in the tanks, it is not necessary to provide for self-cleaning velocities. Usually the design velocity is only 0.3 m/s. The pipes can be as small as 75 mm in diameter, and a continuous downward gradient is not required; a nominal overall gradient (say, 1 in 200) is sufficient. The flat gradients obviate the need both for deep excavation and for pumping, and in flat areas this should allow considerable savings in cost. However, if the cost of the septic tanks is included, the typical construction costs of small-bore and conventional sewerage are roughly similar.

Small-bore sewers may therefore be appropriate:

- where septic tanks already exist, but soakaways have failed or do not exist, so that the effluent runs into street drains;
- where pour-flush toilets are used, but on-site disposal is impossible;
- where sewerage is needed, but the normal conditions of sewer-laying cannot be met without exceptional expense.

Whatever the circumstances, small-bore sewer systems depend on regular and efficient emptying of the septic tanks. Desludging cannot be done at the last minute when the soakaway begins to overflow, or when the owner requests it (which is usually the same thing);

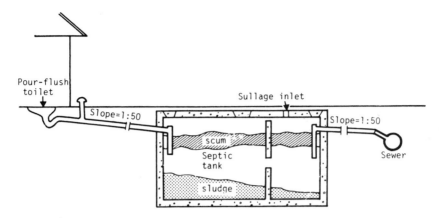

Figure 8.12 Small bore sewerage. In this case the sewer serves a pour-flush toilet with a septic tank
Source: After Kalbermatten, *et al.* (1982)

it must be done routinely at a predetermined interval (say, every 2 years) to ensure that the sewers do not become blocked with solid material.

8.9 OTHER SYSTEMS

The following systems are not generally recommended, but we consider them briefly as they are already in widespread use.

Bucket latrines

One of the oldest and generally least hygienic systems is the bucket latrine. A squatting slab or seat is placed immediately above a bucket which is filled within a few days by the excreta of an average family. The bucket is positioned adjacent to an outside wall and is accessible from the street or lane. A collector calls regularly — preferably every day but more typically once or twice a week — and empties the bucket. The act of emptying the bucket into the barrow or cart typically involves spillage and the area becomes heavily contaminated. The same occurs at the depot where the contents of the carts are emptied for transportation in trucks or for treatment, composting, or agriculture.

Ideally, buckets should be sealed with a lid and carried to the depot before being emptied, a new disinfected bucket being substituted. The practice of emptying the bucket and immediately replacing it should be rejected. At the depot, buckets should be washed thoroughly and painted with a disinfectant. It may be helpful to have a colour code so that all buckets collected on Mondays are red, for instance, and all replacement buckets on that day are yellow.

This will help to distinguish the buckets which have been disinfected from those which have not.

A bucket system can only work well under situations of tight institutional control, where all operations are carefully supervised. It should be regarded as a temporary measure suitable for camps, for instance, while more permanent solutions are being constructed.

Aqua-privies

An aqua-privy is essentially a septic tank located directly underneath a squatting plate which has a 100–150 mm diameter vertical drop-pipe extending some 100 mm below the liquid level in the tank, thus forming a crude water seal.

In practice, maintenance of the water seal, necessary to prevent mosquito and odour nuisance, has proved difficult because a low inflow or a leak in the tank causes the water level to fall. Even discharging sullage into the tank has not proved satisfactory. With relatively high sullage flows and especially in high-density areas, soakaways cannot dispose of the effluent, so that it is discharged into a small-bore sewer, in which case a second compartment is required to settle out the solid matter.

At low sullage flows the aqua-privy is essentially equivalent either to a VIP latrine with a separate soakaway for sullage, or to a pour-flush toilet whose offset soakaway can also receive the sullage. These systems are less expensive than aqua-privies and less prone to malfunction. The pour-flush toilet has a water seal which is much superior to that of the aqua-privy, it does not require a watertight tank, it can be located inside the house and it is more easily upgraded into a cistern-flush toilet. Similarly, when sullage flows are high, the sewered pour-flush system (Figure 8.12) is superior to the sewered aqua-privy.

Cesspools

A cesspool is a covered chamber receiving all waste waters from a dwelling or dwellings. If the cesspool is sealed and has an outlet pipe it is indistinguishable in principle from a septic tank. An unlined cesspool may act as a soakaway (see below) until all pores in the surrounding soil become clogged, whereupon it becomes a septic tank. A more common usage of the word cesspool describes a sealed tank with no overflow in which liquid wastes and sludge are stored. Frequent emptying (at least once every six months) is required and the system is therefore expensive and dependent upon whichever central authority is operating the collection service. The cesspool could then be considered as an expensive variation of the vault system.

Figure 8.13   This septic tank produces about 1000 mosquitoes each night, emerging through the crack in the cover which was caused during desludging. Septic tanks should be thoroughly sealed and covered with soil. Otherwise, they should be treated with polystyrene beads (Section 15.2). If necessary, a floating bar of polystyrene will keep the beads from escaping at the overflow (Photo: S Cairncross)

Compost toilets

In order to encourage the emptying of latrines by householders or others, they may be designed in such a way as to permit composting (Chapter 13) to transform the excreta into a form which can safely be used as fertilizer. The double-vault VIP latrine (Figure 8.5) is in fact a kind of composting latrine. Experiments with other types in which the composting process is accelerated, thus permitting a smaller chamber and more frequent emptying, have shown them to be unsuitable for many developing countries. Although one type has found widespread application in northern Vietnam, there is reason to believe that it has contributed to the spread of *Ascaris* infection there.

Obviously, latrines of this type cannot be introduced into a new area without an enormous supporting effort in communication, education, and evaluation to ensure that they are being properly used. This supporting programme increases the cost of compost toilets.

8.10 COMMUNAL LATRINES

Communal latrines are usually cheaper per capita to build than individual household latrines. But they have many disadvantages and the decision to introduce them should not be taken lightly. The basic problem is that there is little commitment by individual users to keep them clean and operating properly. It is therefore essential to provide one or more well-paid attendants to keep the installation in good order, and lighting and a water supply must also be provided. The latrines should also be regularly inspected by the employers of the attendants, to ensure that they are being well maintained.

Additional problems are:

(i)   the lack of privacy;
(ii)   the difficulty of their use at night and in bad weather, especially by children, the sick, and the old;
(iii) their requirement for public land, which in a crowded shanty town varies between 5 and 10% of the total land area available;
(iv)   they cannot be upgraded to individual household latrines.

The ideal type of toilet to install is a pour-flush or low-volume cistern-flush toilet, at the rate of one compartment for every twenty-five people served. If shower and clothes-washing facilities are not available in individual households, these should also be provided. Figure 8.14 shows a layout for a communal sanitation facility serving about 300 people.

8.11 SOIL CONDITIONS

It is more difficult to excavate pits and pipe trenches in areas of rock or loose sand, and where there is a danger of ground water contamination this of course influences the choice of sanitation system. In addition, the feasibility of some systems for on-site disposal of excreta is limited by insufficient infiltration capacity in the local soils. No large-scale sanitation programme should therefore be undertaken without a thorough soil survey. Ground water contamination has been discussed in Section 8.4. Infiltration capacity can be estimated as follows.

The infiltration capacity of a soil depends largely on the size of its particles. Sand particles are defined as those larger than 0.05 mm in diameter, clay particles are those finer than 0.002 mm, and those between these two sizes are known as silt. The soils with the highest infiltration capacity are sands, with more than 80% sand particles by weight. If rubbed in the palm of the hand, they will not stain the skin. Somewhat less permeable are silts and loams, which have a mixture of particle sizes. The least permeable are clays, with more than about 35% clay particles. They can usually be distinguished from other soils by the following test.

Prepare sufficient soil to form a ball about 20 mm in diameter by moistening with water. Place the prepared sample in the palm of one hand, and shake vigorously by jarring the hand on a firm object or against the other hand. Continue shaking until water comes

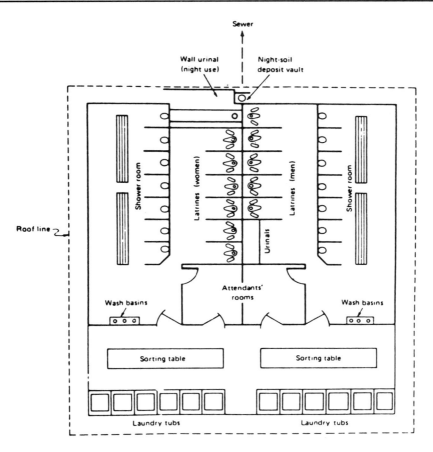

Figure 8.14   Layout of a communal latrine and washing facility serving about 300 people
Source: From Kalbermatten *et al.* (1982)

to the surface of the sample producing a smooth shiny ('livery') appearance. Squeeze the sample between thumb and forefinger of the other hand, and the surface water will disappear so that the surface becomes dull. However, if the sample is of a clay neither of these changes in surface texture will occur. Where clay soils are detected, more detailed site investigation is especially important, including percolation tests, to establish the real infiltration capacity of the ground.

Approximate rates of infiltration for sewage and for sullage in three types of soil are given in Table 8.1. The infiltration rate for sewage (in a soakaway from a pour-flush toilet, for instance) is lower because the sides of the pit tend to become partially clogged with organic matter after a few months' operation. The infiltration rate for settled effluent from a modified septic tank (Section 8.7) would be somewhere between the rates for sewage and sullage.

From Table 8.1, and the load of liquid matter to be infiltrated, it

Table 8.1   Approximate infiltration capacities of various soils

Soil type	Infiltration capacity (l/m^2.day) Sewage	Sullage
Sands	50	200
Silts and loams	30	100
Clays	10 or less	50 or less

is possible to estimate the surface area required on the sides of a pit or soakaway trench below the highest permissible water level. The base of the pit or trench cannot be counted as infiltration area, because it is rapidly blocked by sludge.

A typical load on a pit latrine used by five to ten people is 8 to 15 l/day. Where water is used for anal cleansing, this would increase to 28–55 l/day. In a conventional pour-flush toilet, it would be 50–100 l/day. Where a soakaway pit or trench is also to receive sullage, the load is roughly equal to the volume of water used daily in the house, and therefore depends on the level of water supply. Typical volumes of sullage water production are:

communal wells and hand pumps   10 l/person.day
communal standpipes   15 l/person.day
household wells, yard taps   30–50 l/person.day
multiple-tap private connections   50–300 l/person.day.

8.12 SOAKAWAY DESIGN

Figure 8.8 illustrates the construction of a soakaway pit. The upper 300 mm of the walls are made impermeable, and the cover slab can be buried to keep out insects. In less permeable soils, care should be taken to avoid smearing the sides while excavating the pit, as this can reduce their permeability. The sides of the pit should be scored with the edge of a spade as the lining is built, to improve their infiltration capacity. It will also help to place a layer of fine gravel or sand between the lining and the sides of the pit.

If the area required for infiltration implies an uneconomically large pit, soakaway trenches may be used instead. These are filled-in trenches containing open-jointed pipes of 100 mm diameter, laid on rock fill, gravel, or broken bricks. They provide a larger infiltration area for a smaller volume of excavation than a pit. They should be deep and narrow to obtain the greatest possible side-wall infiltration area.

Normally several trenches are dug, each about 15–30 m long, and connected together to make a drainfield. The trenches should not operate in parallel through distribution boxes, but in series as shown in Figure 8.15 so than as each trench fills, it overflows to the

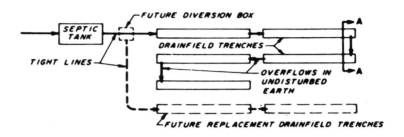

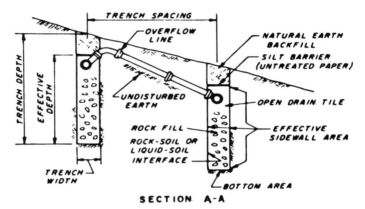

Figure 8.15 Typical drainfield arrangement for septic tank effluent *Source*: From Cotteral and Norris (1969). Reproduced by permission of the American Society of Civil Engineers

next one. This ensures that each trench is used either fully or not at all, avoiding the formation of an impermeable crust on the sides.

The minimum spacing between trenches should be 2 m or twice the trench depth, whichever is greater, and each trench should be level with, or below, the ones before it. An equal area of land should be kept in reserve for possible extension or replacement of the drainfield if it becomes clogged in the future. The total length of trench required is calculated from the equation:

$$L = \frac{NQ}{2DI}$$

where L = trench length in metres;
        N = number of users;
        Q = wastewater flow in litres per capita daily;
        D = effective depth in metres;
        I = infiltration rate in l/m^2.day (Table 8.1)

The factor 2 is introduced because the trench has two sides.

Where there is a risk of ground water contamination, or where the soil is impermeable or difficult to excavate, a possible solution is the soakaway mound (Figure 8.16). This ensures a greater depth and dispersion of the effluent's travel into the soil, as well as removing much of its water content through the evapotranspiration of the grass planted on the top.

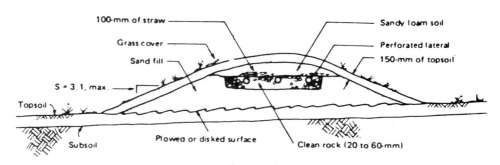

100-mm of straw — Sandy loam soil
Grass cover — Perforated lateral
Sand fill — 150-mm of topsoil
S = 3:1, max. —
Topsoil —
Subsoil — Plowed or disked surface — Clean rock (20 to 60-mm)

Cross section a-a

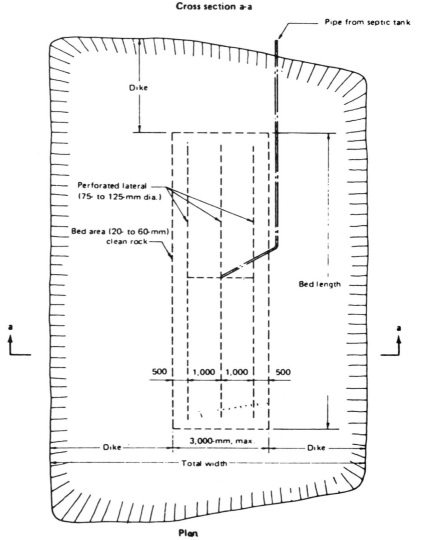

Pipe from septic tank

Dike

Perforated lateral
(75- to 125-mm dia.)

Bed area (20- to 60-mm)
clean rock

Bed length

a   a

500 | 1,000 | 1,000 | 500

3,000-mm, max.

Dike — Total width — Dike

Plan

8.13 PERSONAL HYGIENE

Only limited benefits will be achieved from well-designed, well-constructed, and well-maintained latrines unless people wash their hands after using them, particularly where water is used for anal cleansing. Provision for hand-washing should be an integral part of any sanitation programme, whether in the form of a simple clay water pot, a tap in the yard near to the latrine or, if possible, a wash basin. The hand-washing facility should be accessible to children. In those parts of the world where sand is often used for washing, allowances should be make for the possibility of this filling up septic tanks prematurely.

8.14 SULLAGE DISPOSAL

Sullage is domestic waste water not containing excreta or toilet wastes. It is sometimes called grey water in distinction to sewage or black water. Sullage always contains some pathogens, often at a concentration similar to that in sewage. Pools of sullage on the ground are a breeding place for *Culex pipiens* and some *Anopheles* mosquitoes (Chapter 15), and sullage thus poses a health risk.

Sullage may be disposed of by special sullage soakaways if the soil conditions are suitable, by stormwater drains, or by sewers if these are available (for example as part of a sewered pour-flush system). Some ingenuity on the part of design engineers is required to provide low-cost sullage disposal facilities; for example if sullage is to be discharged into stormwater drains, the drains may need to be of a cross-section that permits the sullage to flow at a reasonable velocity in the dry season (Chapter 11). Treatment of the sullage may be necessary to prevent gross pollution of the receiving watercourse at these times: one solution might be a facultative waste stabilization pond (see Chapter 10), but to protect the pond from overflowing in the rainy season if would be necessary to install a stormwater overflow weir before the inlet.

8.15 NIGHTSOIL AND SLUDGE RE-USE AND DISPOSAL

Human wastes are a valuable natural resource and should be re-used wherever practicable. There are essentially three different systems for re-use; agriculture, aquaculture, and biogas production. A brief discussion of each method is presented in order to make the reader

Figure 8.16   Prototype design for an evapotranspiration mound, for use in areas where the water table or rock is near the surface (S=slope, dimensions are in millimetres)
Source: From Kalbermatten *et al.* (1982)

aware of the possibilities. The health hazards associated with excreta re-use are discussed in Chapter 14.

Agriculture

Both nightsoil and sludge may be used to enrich the soil. The direct application of nightsoil as an agricultural fertilizer has been practised for centuries in many parts of the world, but this practice involves very substantial health hazards to agricultural workers and to the consumers of the crops grown, and is not recommended. It is particularly inadvisable on vegetables or any other crop which may be eaten uncooked. Instead, nightsoil or sludge should be either digested, or composted with organic refuse and vegetable matter (Chapter 13), before application to the land. Chapter 14 further describes the health hazards of agricultural re-use.

Aquaculture

Ponds containing human wastes are rich in aquatic life provided they remain in an aerobic condition. In particular, they support large blooms of algae. These conditions are perfect for the growth of certain types of fish, especially carp and tilapia. Fish productivities of over 3 t/ha.year can be achieved in ponds enriched with excreta. Ducks also thrive, in conjunction with fish, in such ponds.

It is possible to feed fish ponds with nightsoil brought in carts from a nearby community, or to locate latrines over ponds for the direct entry of excreta, or to grow fish in the maturation ponds in a series of waste stabilization ponds which are treating water-borne wastes (see Chapter 10). All these arrangements are commonly employed today, particularly in South East Asia.

The fish harvested may be used to produce high-protein meal for pig or poultry farming but more usually are sold for domestic consumption. However, if fish are to be sold for human consumption it is important that they are well cooked. If not, they may act as the main transmission route for a variety of pathogens, most notably the oriental liver fluke (*Clonorchis sinensis*). Further details of this problem are given in Section 14.3.

Biogas

Nightsoil, mixed with animal wastes, can be used to generate methane, or 'biogas'. This technique has been applied in India, Korea, and in China, where nearly one million biogas plants are in use. In a warm climate, a biogas plant in which a family's excreta are supplemented with the dung from several animals, such as cattle or pigs, can meet the fuel needs for cooking and lighting of a family of five people.

Figure 8.17 shows a typical biogas plant for one family. The influent human and animal waste is mixed with straw or other crop wastes and diluted with an equal volume of water. As this 'slurry' is fed into the digester, the effluent or spent slurry is removed for fertilizer and the gas produced by anaerobic digestion accumulates in the space above.

Biogas production requires a warm climate; about half as much is produced at 25°C as at 30°C, and almost none at all below 15°C. It also requires an initial investment too large for many rural families. A biogas plant requires careful control and is only appropriate where good maintenance can be guaranteed. It is necessary to clean out most of the accumulated sludge at least once a year, to keep the pH between 7 and 8, and to ensure correct mixing and dilution of the slurry. The slurry may also require occasional stirring. In countries with a low population density, as in many parts of Africa, where each person's productivity is at a higher premium than the productivity of any given hectare of land, it may be felt that the production of gas and fertilizer do not justify this heavy investment of labour.

Biogas can be generated from nightsoil on an individual family scale or at a large central facility. Biogas can also be generated from the sludge produced at a conventional sewage treatment works. A comprehensive review of this subject has been prepared by Barnett *et al.* (1978).

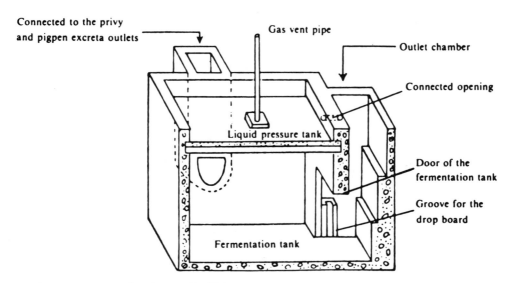

Figure 8.17   A biogas digester used in China
Source: From McGarry and Stainforth (1978). Reproduced by permission of the International Development Research Centre

Disposal

Where it is not feasible to re-use untreated nightsoil or sludge, it may be added to the inflow to the town's sewage works. However, nightsoil and sludge from vault, bucket and pit latrines often contain large amounts of domestic refuse and bulky anal-cleansing materials, and special alterations to the works may be necessary.

As a last resort, the material can be buried in trenches at least 0.6 m deep and 0.6 m apart. The nightsoil or sludge is placed in them to not less than 0.3 m below ground level, covered with tamped earth to make a small mound of earth over the trench, and then left for at least two years.

The chief advantage of the method is simplicity, although good supervision is required to ensure that the trenches are not filled too full of nightsoil. The disadvantages include the cost of land required, the possible contamination of ground water for water supplies and the depredations of wild or domestic animals. Some simple form of re-use such as composting is often preferable to this.

8.16 WATER AVAILABILITY

Several of the technology options for sanitation depend on water for flushing, whether by pouring or from a cistern, and so the feasibility of these options depends on the reliability and level of service of the water supply. On the other hand, the pattern of domestic water use influences the most appropriate method of waste water disposal. It is particularly important to study water use in the case of properties with individual water supply connections, because the waste water flows may be too large for some of the non-sewered systems.

Even with multiple taps per property, it may be possible to reduce domestic water consumption to 100 l/person.day or less, by the use of low-volume cistern-flush toilets and various simple devices for reducing the rate of water flow from taps and showerheads (Chapter 6). Cistern-flush toilets account for 30–50% of domestic consumption, but can be designed to reduce the volume of flush water from 10–20 l/flush to 3–6 l/flush.

With reduced domestic water consumption, the limitations of non-sewered sanitation systems need not conflict with the desirability of a high level of water supply service. Sewers are required to dispose of large volumes of sullage and flushing water, not excreta, so that the elimination or reduction of non-essential water use is a key element in an economic solution to sanitation problems.

8.17 REFERENCES AND FURTHER READING

Barnett, A., Pyle, L. and Subramanian, S.K. (1978). *Biogas Technology in the Third World: a Multidisciplinary Review* (Ottawa: International Development Research Centre).

Brandberg, B. (1983). *The Latrine Project, Mozambique*, Report No. IDRC-MR58e (Ottawa: International Development Research Centre).

Cairncross, S. (1988). *Small Scale Sanitation*, Ross Institute Bulletin No. 8 (London: London School of Hygiene & Tropical Medicine).

Carroll, R.F. (1980). *Affordable Sanitation for Developing Countries*, Note No. N147/80 (Garston, England: Building Research Establishment).

Cotteral, J.A. and Norris, D.P. (1969). Septic tank systems, *Journal of the Sanitary Engineering Division, ASCE*, **95**, 715–746.

Curtis, C.F. and Feachem, R.G. (1981). Sanitation and *Culex pipiens* mosquitoes: a brief review, *Journal of Tropical Medicine and Hygiene*, **84**, 17–25.

Curtis, C.F and Hawkins, P.M. (1982). Entomological studies of on-site sanitation systems in Botswana and Tanzania, *Transactions of the Royal Society of Tropical Medicine and Hygiene*, **76**, 99–108.

Kalbermatten, J.G., Julius, D.S., Gunnerson, C.G. and Mara, D.D. (1982). *Appropriate Sanitation Alternatives: a Planning and Design Manual* (Baltimore: Johns Hopkins University Press).

Lewis, J.W., Foster, S.S.D. and Drasar B.S. (1982). *The Risk of Groundwater Pollution by On-site Sanitation in Developing Countries; a Literature Review*, IRCWD Report on 01/82. (Duebendorf, Switzerland: International Reference Centre for Wastes Disposal).

McGarry, M.G. and Stainforth, J. (1978). *Compost, Fertilizer and Biogas Production from Human and Farm Wastes in the People's Republic of China* (Ottawa: International Development Research Centre).

Mara, D.D. (1984). *The Design of Ventilated Improved Pit Latrines*, TAG Technical Note No. 13 (Washington DC: World Bank).

Mara, D.D. (1985). *The Design of Pour-Flush Latrines*, TAG Technical Note No. 15 (Washington DC: World Bank).

Mara, D.D. and Sinnatamby, S.G. (1986). The rational design of septic tanks in warm climates. *The Public Health Engineer*, **14**(4), 49–55.

Morgan, P.R. (1977). The pit latrine revived, *Central African Journal of Medicine*, **23**, 1–4.

Morgan, P.R. and Mara, D.D. (1982). *Ventilated Improved Pit Latrines: Recent Developments in Zimbabwe*, TAG Working Paper No. 2 (Washington, DC: World Bank).

Okun, D.A. and Ponghis, G. (1975). *Community Wastewater Collection and Disposal* (Geneva: World Health Organization).

Otis, R.J. and Mara, D.D. (1985). *The Design of Small-Bore Sewer Systems*, TAG Technical Note No. 14 (Washington DC: World Bank).

Pacey, A. (1978). *Sanitation in Developing Countries* (London: John Wiley).

Pacey, A. (1980). *Rural Sanitation: Planning and Appraisal* (London: Intermediate Technology Publications).

UNCHS (Habitat) (1986). *The Design of Shallow Sewer Systems*, (Nairobi: UN Centre for Human Settlements).

Van Buren, A. (1979). *A Chinese Biogas Manual* (London: Intermediate Technology Publications).

Wagner, E.G. and Lanoix, J.P. (1958). *Excreta Disposal for Rural Areas and Small Communities* (Geneva: World Health Organization).

Winblad, U. and Kilama, W. (1985). *Sanitation without Water* (London: Macmillan).

9

Planning a Sanitation Programme

9.1 PROGRAMME CONTEXT AND CONTENT

'A sanitation programme' is taken in this chapter to refer to a programme for the improvement of excreta disposal facilities. We are concerned here with sanitation programmes for low and middle-income communities. A more detailed treatment of the subject of this chapter is in Cairncross (1992).

Context

Broadly speaking, sanitation programmes in developing countries may be carried out in three contexts.

(1) Site-and-service schemes, where new housing and infrastructure are to be installed. This case poses least administrative problems as the excreta disposal facilities may be built directly by a centralized authority such as a municipality, housing follows a planned layout, and the cost of construction may be recovered from householders by the process of site purchase or rental.

(2) Shanty town or squatter upgrading schemes, where an important planning decision is in the balance of responsibility between the sanitation agency (usually the municipal authority) and the individual household. Where the households are expected to build their own toilets in a 'self-build' scheme or to pay an installation or connection fee, it will not normally be possible to cover the whole community in the short term as some households will opt out of the scheme. If, therefore, it is desired to improve sanitation rapidly for all families in the area, the agency will have to finance the scheme and recover the cost by other means, such as local taxes or a surcharge on the water rates.

In upgrading schemes, the existing housing configuration and land tenure pattern often make the installation of infrastructure such as sewers or communal latrines more difficult, although local organization may be stronger, and thus facilitate some form of community participation. Squatters require some guarantee of security against eviction, if they are to invest time or money in sanitation improvements.

A sanitation programme in this context usually benefits from inclusion in a general programme for infrastructural upgrading, including roads, drainage, and water supply, as this can more easily guarantee the necessary funding, land, water, street alignment, and recovery from the households of the money invested. It can also improve general hygiene and provide the necessary conditions for a change in excreta disposal practices, thus rendering more likely the health benefits which are hoped for from the programme.

(3) Rural sanitation programmes generally rely heavily on householders to build their own latrines, although they may be given some assistance, for example by the sale or distribution of slabs for pit latrines. Rural programmes therefore require the greatest input of health education, technical assistance, and follow-up extension work if they are to succeed. Local organizations can play an important role. For example, the councils of some new communal villages in Mozambique passed a bye-law that each household must complete a pit latrine before starting to build a new house.

Content

Whatever the context, there are certain key elements which a sanitation programme should include:

  (i)   a central steering committee comprising the ministries or departments responsible for finance and planning, health, urban or rural development, water supply and sewerage;
  (ii)   sound project management, site investigations, careful technology choice and design;
  (iii) pre-programme study of social factors, economic constraints, and beneficiary preference;
  (iv)   development of an extension system, including health education, technical assistance to self-building householders, and feedback from the community;
  (v)   access to and delivery of building materials and mass-produced components (Figure 9.1), combined with financing mechanisms;
  (vi)   integration of designs with infrastructure development, particularly water supply, stormwater drainage and housing layouts;
  (vii) integration of programme management with existing administrative structures, such as village or town councils;
  (viii) a monitoring and evaluation programme;
  (ix)   a programme for briefing central government personnel, and

for training engineers, technicians, artisans, and extension workers.

Programmes involving systems for on-site disposal of excreta, and especially those with a significant self-building component, need different institutional arrangements from those appropriate for sewerage (Chapter 7). They require more decentralization and links with local community bodies, and more co-ordination with health, housing and other agencies such as housing banks.

For the most beneficial results, a sanitation programme should cover all households in the community; there is little benefit to community health, for instance, if only the richer or more conscientious households use latrines while the others foul the streets. Programmes depending on individual household choice will take a very long time to achieve this, if they ever do. In 'self-build' programmes therefore, there is some advantage in making the decision to build latrines a collective one, possibly by financing them for groups of families at a time. On the other hand, groups of households will have difficulty with the regular collection of funds for repayment of a collective loan. The same applies to the payment of maintenance bills for communal latrines. If collective arrangements are made for payment of costs, then special dispositions are required for late entrants such as families who initially opted out or could not pay, and those moving into the area.

9.2 CHOICE OF SYSTEM

Health benefits from improvements in excreta disposal are difficult to detect and even harder to quantify, but from what is known (see Chapter 1), we can deduce that there is little difference between the various types of system so far as public health is concerned, as long as they are correctly built, maintained and *used*. The principal objective of the conventional cistern-flush system is to provide a higher level of convenience, not better health.

Other things being equal, then, the selection of one design over another is determined by a composite of technical, social, and economic factors. Some of these factors are illustrated in Table 9.1.

The process of choice should begin with an examination of all the options, described in Chapter 8. There are usually some which can readily be excluded for technical or social reasons. Typical technical reasons are water availability and soil conditions (Chapter 8), particularly in areas of high population density. An example of a social reason is the unsuitability of pour-flush toilets for people who use large objects for anal cleansing. Once these exclusions have been made, cost estimates are prepared for the remaining technologies, and some can be quickly excluded on grounds of excessive cost.

a

b

Since the benefits of the various sanitation systems cannot be quantified, the final step in identifying the most appropriate type must rest with the intended beneficiaries. Those alternatives which have survived technical, social and economic tests are best presented to the community with their attached price tags, so that the users may decide what they are able and willing to pay for. The users should thus be able to choose from a range of types, suitable for different groups or neighbourhoods with varying levels of income, water supply, and housing standards.

Figure 9.2 illustrates how the technical, social, and economic checks may be co-ordinated in practice. The principal technical factors — soil conditions, population density, and water availability — have been discussed in Chapter 8. The social and economic factors are considered below.

9.3 SOCIAL FACTORS

In planning and development, the 'social factor' is often invoked to explain the failure of technically sound projects. It is suggested that there is something strange and exotic in the community's values or institutions which can account for their lack of co-operation. But there is another side to these social problems; planners and administrators bring their own values into programmes, often without realizing that they are not shared by the community.

Thus, while there may be some *exotic* factors arising form the particular culture of the users of a sanitation system, there is a range of *mundane* difficulties which arise from differences in viewpoint between the planners and the users, and which are easier to predict, analyse, and remedy, largely by common sense and a knowledge of how the users live.

An example of an exotic factor found in various countries is that sons-in-law and mothers-in-law avoid using the same latrine. Here it may be easier to provide a spare latrine than to try to persuade them to change their ways. Another such factor is the taboo in some Muslim countries against defecating while facing Mecca. The implications for latrine construction are obvious.

The mundane factors which are sometimes not considered are the cost, convenience, and organization necessary for the users if they are to make correct use of the sanitation system.

The cost of a new toilet may compete as a priority with extra living space or other home improvements, and some systems may

Figure 9.1 Casting pit-latrine squatting slabs (a) in Barbados, and (b) in Mozambique, where conical slabs (Figure 8.2) are used. Both urban and rural latrine programmes can benefit greatly from the mass production and distribution of essential latrine components, such as squatting slabs, vent pipes, and pour-flush pans and traps
(Photos: (a) E Rice, WHO, (b) B Brandberg)

Table 9.1　Comparison of several types of sanitation system

Sanitation system	Rural application	Urban application	Construction cost	Operating cost	Ease of construction	Water requirement	Soil conditions required
Pit latrines	Suitable	Not in high-density areas	Low	Low	Very easy except in wet or rocky ground	None	Stable permeable soil; water table >1 m deep
Pour-flush toilets	Suitable	Not in high-density areas	Medium	Low	Requires builder	Water near toilet	Permeable soil; water table >1 m deep
Sewered pour-flush toilets	Not suitable	Suitable	High	Medium	Requires engineer	Water piped to house	Preferably stable soil; no rock
Vault toilets and vacuum trucks	Not suitable	Suitable where vehicle access and maintenance available	Medium	Very high	Requires builder	None	None
Septic tanks and soakaways	Suitable	Suitable in low-density areas	High	High	Requires builder	Water piped to toilet	Permeable soil; water table >1 m deep
Conventional sewerage	Not suitable	Suitable where affordable	Very high	High	Requires engineer	Water piped to toilet	Preferably stable soil; no rock

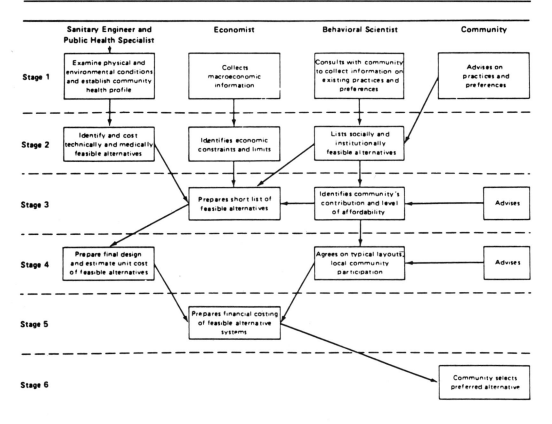

Figure 9.2 Recommended structure for feasibility studies for a sanitation programme
Source: From Kalbermatten *et al.* (1982)

require new payments, such as water bills, for their operation or their upkeep. A sanitation system may appear to be gaining quite wide acceptance without reaching the poorest people, who most need it. Cost is further discussed below.

Convenience is a rather wide concept, but includes such factors as how much time is saved, how the location of facilities influences their use, and whether facilities are equally convenient to all household members and at night. It is particularly important that children should have easy access to toilets and that there should be no fear, justified or not, associated with their using them (Figure 9.3). In the case of self-build schemes, an important factor is the ease of obtaining building materials and prefabricated components such as vent pipes or squatting pans.

Organization becomes critical when toilets are to be used communally. In general, if facilities must be shared between households, problems will arise unless someone is appointed and adequately rewarded to look after them and keep them clean. Even in Ibadan,

Figure 9.3 This scene from El Salvador illustrates the importance of latrine use by children. Children under 10 years-old are the main sufferers from most of the diseases listed in Table 1.3. They are also the main excreters of the pathogens that cause these diseases. Therefore the health of the whole family depends upon the use of latrines by children old enough, and the thorough cleaning away by the mother of the stools of babies. The family latrine must be kept clean to protect the health of children who use it (Photo: WHO)

Nigeria, with strong ethnic and family ties between neighbours and a rural tradition of communal latrines, it has not proved possible to collect regular cash contributions for the maintenance of communal urban latrines (Pasteur, 1979). The local authority cannot abdicate its responsibility to ensure that communal latrines are satisfactorily operated and maintained. If it is considered essential to recover the operating and maintenance costs directly from the users, it will be more reliable to charge for admission. This has been successfully tried in Ghana and also in India, where the small admission fee also covers the cost of soap for washing. However, a charge may discourage some people, especially children, from using the latrines. In some areas, free latrines for children are provided next to the public ones.

The likely effects of the various social factors on a sanitation programme may be appraised to some extent by studying existing excreta disposal practices. Still more valuable is a study of similar communities where new sanitation technology has been introduced and accepted — or rejected. A look at the records will show how many users are contributing to maintenance, and how much. Observation of how facilities are being used or misused will suggest some design modifications, and others will be found to have been carried out already by individual householders. Local staff and community representatives can point to problems which arose during construction, such as shortages of materials, over-rigorous building regulations, insufficient supervision of construction, and so on.

The extension system

It is the function of the extension system to find and deal with any problems of this kind which have not been resolved at the planning stage. An extension system is particularly important where self-building is involved, and should promote the construction, correct use, and maintenance of improved toilets through health education and other means, while at the same time collecting feedback from the community about difficulties and design improvements.

Several guides are available to the integration of health education in sanitation programmes (Perret, 1983, 1984; IRC, 1991). Community-level primary health care workers are suitable promoters for a sanitation programme, provided they are given the necessary training, and supported through visual aid materials, radio campaigns, and the like. Other means may include technical assistance, demonstration models, the installation of latrines of the appropriate type in schools (Figure 9.4), clinics, and community centres, and the advertising of credit arrangements and other forms of help to the householders.

The extension system will itself cost money and require staff, which must be provided for, in good time, and these costs should be taken into account when comparing the costs of alternative systems.

9.4 COSTS

It has been pointed out in Chapter 7 that whether or not a loan is obtained to finance a sanitation programme, the community must eventually pay for it, and it must therefore be affordable. There are limits to how far sanitation for the poor can be subsidized by the rich, though this is possible to a certain extent through progressive charging systems (Chapter 6). To the extent that users will be expected to pay all or part of the cost, therefore, it is important to know how much they can pay.

It has been estimated by the World Bank that housing typically represents 15 to 20% of household expenditure in developing

Figure 9.4 School latrines in Guyana. The provision of well-designed and well-maintained latrines in schools, combined with classroom education on hygiene and disease, is a crucial component of any sanitation programme. Unfortunately, it is common to see school latrines that are dirty and not used by many children, thus providing a negative message for the children to take home
(Photo: E Rice, WHO)

countries, and the lower figure is probably more appropriate for low-income groups. Detailed studies in Tanzania, Egypt, and elsewhere indicate that expenditure on infrastructure ranges between 50 and 60% of this total, and that sanitation accounts for a quarter to a third of the infrastructure expense. It may thus be estimated that low- and middle-income groups will typically be able to afford to spend only 2–3% of their income on sanitation, although they may be prepared to spend larger sums over a short period of time.

However, it is not very reliable to estimate the householders' willingness to pay for sanitation improvements on the basis of their earnings levels. It is preferable to ask a sample of people outright what they would be prepared to pay or, better still, to implement the programme on a pilot scale and evaluate the response. If only a few people are prepared to pay for the improvement, a second survey may be used to investigate the reasons for this.

The total annual cost per household of any given sanitation system may be obtained by writing off the construction cost over a reasonable period of time (such as five years for a low-cost system) and at an appropriate discount rate (such as 10%), and adding to this the annual cost of maintenance, flushing water, pit-emptying, nightsoil or sewage treatment and disposal, etc. This cost may then be compared with what the beneficiaries are able and willing to spend, and the funds at the sanitation agency's disposal. The result will usually be a considerable narrowing of the range of feasible technologies. The calculation of costs can be done in two different ways, to derive economic and financial costings respectively.

Economic costs

Economic costing is used by national planners and policy-makers to make a least-cost comparison between alternatives, and includes all costs to the economy regardless of who incurs them, but does not include taxes and eliminates price distortions created by financial legislation. In an economic cost calculation, cost components in foreign exchange may be increased by a certain percentage, to represent the scarcity of hard currency, and labour costs may be valued at a 'shadow' wage rate lower than the legal minimum wage, representing the fact that there is little real cost to the economy if unemployed labourers are given some work to do. Further details of economic costing methods are given by Squire and Van der Tak (1975), although it is advisable to obtain the help of an economist in applying them.

An economic cost comparison will serve to accentuate the advantages of non-sewered systems, by comparison with conventional sewerage, since a relatively higher proportion of their costs are on labour, and less in foreign exchange. International donor agencies are particularly interested in such arguments.

The wide variation in costs between different systems is illustrated by the results of a survey conducted by the World Bank in several developing countries, which found that the mean annual economic costs per household of various sanitation technologies, relative to conventional sewerage, were as follows:

improved pit latrines	10
pour-flush toilets	10
sewered pour-flush toilets	40
vault toilets	50
conventional septic tanks	90
conventional sewerage	100.

These figures should not, however, be applied to specific settings; the costs of the various options must be calculated afresh for each specific context.

An economic cost comparison can be useful for showing up distortions which are unnecessarily favourable to one or other type of system, but which can be corrected by changes in government policy. One example of such a distortion would be the subsidies or grants provided in many developing countries by the central government for the construction of conventional sewerage and sewage treatment systems; it could be remedied by also making funds available for other types of system. Another example is the charging of water rates below cost to large domestic consumers, which encourages the use of systems wasteful of water; large individual consumers should be charged at least the real cost of their water, and under a progressive tariff system (Section 6.6) would pay still more.

Financial costs

Economic cost comparisons are not normally used for choosing the technology to be used in specific sanitation programmes, as the costs are in practice usually borne by penurious local authorities and by the beneficiary households. It is of little interest to them to know, for instance, that one or other type of toilet would be cheaper if they did not have to pay purchase tax on cement. They are more interested in a financial cost estimate, which is concerned with the actual payments to be made by the various parties concerned — chiefly the sanitation agency and the householder.

It is important to make correct comparisons among true alternatives. For example, some technologies, such as sewered pour-flush latrines (Figure 8.12), dispose of excreta and sullage. By contrast other systems, such as VIP latrines (Figures 8.3 and 8.5) dispose only of excreta. Therefore, in making the comparison between these options the cost of the sewered pour-flush system must be compared with the cost of VIP latrines *plus* a separate sullage disposal system, such as individual yard soakaways. The costs of any waterborne sanitation system must include the costs of flushing water required.

Once the costings for each feasible option have been estimated, consideration should be given to the question of who will pay for each component, and how. In practice the share of construction costs borne directly by the household tends to range from over 90% for some on-site disposal systems to about 50% for conventional sewerage, if the householder pays for internal plumbing. However, there is nothing inherent in the technology which determines these percentages; rather, they should be determined by an explicit policy decision. It might be argued that those installing the cheapest type of system should pay the smallest percentage of the installation cost.

Once it has been decided how much the householder is to pay, the balance of the necessary investments mush be met by a grant or subsidy of some kind. The householder may be assisted to pay his own share of the cost by some form of credit, whether offered to individuals or to groups of neighbours.

9.5 OPERATION AND MAINTENANCE

In the same way as for the construction of excreta disposal facilities, thought must be given to the division of responsibilities between the sanitation agency and the users with regard to their maintenance. If a task such as the emptying of double-pit VIP latrines is to fall to the agency, then the necessary resources must be provided for this purpose. If it is to be the responsibility of the user, then incentives, assistance, and even sanctions may be required to ensure it is carried out, and the extension system (Section 9.3) must become permanent.

9.6 INCREMENTAL SANITATION

The sanitation systems in use in the industrialized countries evolved over more than a century of successive improvements. For example, the collection of nightsoil from bucket latrines in eighteenth-century London was a first step towards reducing gross urban pollution. It was followed by piped water supplies and the development of combined sanitary and stormwater sewerage, and then by separate sanitary sewerage and sewage treatment.

It would be simplistic to think that sanitation in developing countries should be upgraded in a single step, and most unjust to the large urban populations who cannot afford the improvements which in the long term would be most desirable. It is clearly preferable to start with a modest improvement and upgrade it progressively over a period of years or decades.

A major impact on a community's health can be achieved initially by the provision of standpipes and VIP latrines, for example. In the years to come and as the economic status of the community improves, the water supply and sanitation facilities can be upgraded, for example to yard taps and pour-flush toilets initially, and later to a multiple-tap in-house water supply and a sewered pour-flush toilet system. The upgrading could be carried out either by the sanitation agency, or by individuals over a period of time according to their means. This example follows one upgrading route; others are shown in Figure 9.5.

It is noteworthy that none of the upgrading sequences in Figure 9.5 leads to conventional sewerage. This is not because conventional sewerage schemes should not be built (they are a

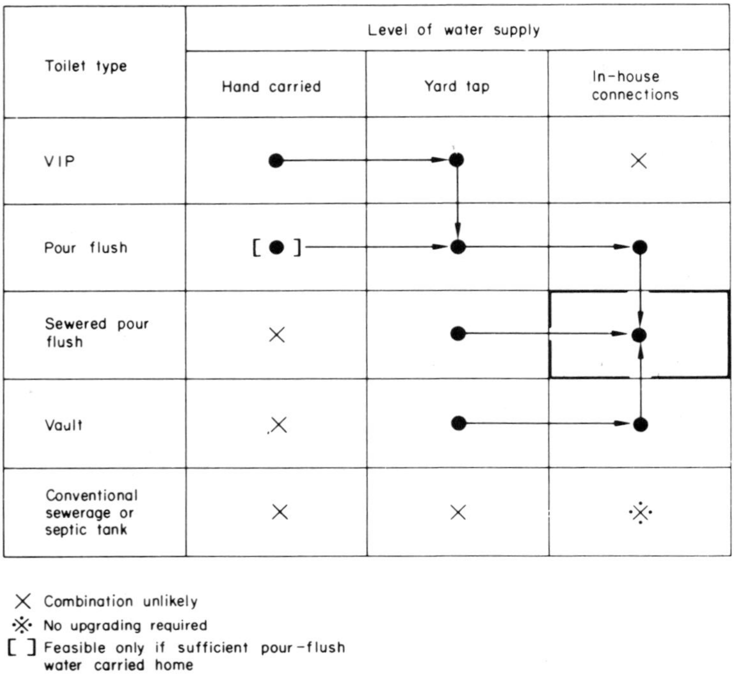

Toilet type	Level of water supply		
	Hand carried	Yard tap	In-house connections
VIP	●————————→●		×
Pour flush	[●]————————→●		●
Sewered pour flush	×	●	●
Vault	×	●————————→●	
Conventional sewerage or septic tank	×	×	⋰⋱

× Combination unlikely
⋰⋱ No upgrading required
[ ] Feasible only if sufficient pour-flush water carried home

Figure 9.5   Potential sanitation upgrading sequences
Source: From Kalbermatten *et al.* (1982)

Table 9.2   Comparative costs of typical planned sanitation sequences

Sequence	Relative economic cost per household per year over a 30-year period

A. Years 1–10; VIP
   Years 11–20; PF
   Years 21–30; SPF
	23

B. Years 1–10; VIP Years 11–30; SPF	42
C. Years 1–30; SPF	51
D. Years 1–30; CS	100

VIP: ventilated improved pit latrine; PF: pour-flush toilet; SPF: sewered PF; CS: conventional sewerage. For example, in Sequence A, a VIP is installed in year 1 and changed to a PF in year 11; the PF is then changed to a SPF in year 21; corresponding changes in water supply would be standpipes to yard taps to multiple in-house connections.
Source: From Kalbermatten *et al.* (1982)

good form of sanitation for those who can afford them and have plenty of water), but because they are not necessary to provide the highest standard of sanitation. The sewered pour-flush system, which can eventually include a low-volume cistern-flush toilet for added user convenience, has an equally high standard of hygiene, and has two big advantages over conventional sewerage; it is substantially cheaper and it can be reached by a staged improvement of several different sanitation technologies. Thus sanitation programme planners can confidently select one of these 'base line' technologies in the knowledge that, as economic status and sullage flows increase, it can be upgraded in a planned sequence of incremental improvements to a sophisticated 'final' solution.

Incremental sanitation sequences are much more cost-effective than conventional sewerage, as is shown in Table 9.2. Many more people can thus be provided with satisfactory excreta disposal facilities for the same amount of money, and these facilities can be upgraded as more money becomes available in the future.

9.7 REFERENCES AND FURTHER READING

Cairncross, S. (1988). *Small Scale Sanitation*, Ross Institute Bulletin No. 8 (London: London School of Hygiene & Tropical Medicine).

Cairncross, S. (1992). *Sanitation and Water Supply: Practical Lessons from the Decade*, Water and Sanitation Discussion Paper No. 9 (Washington DC: World Bank).

Feachem, R.G. (1980). Community participation in appropriate water supply and sanitation technologies: the mythology for the Decade. *Proceedings of the Royal Society of London, B*, **209**, 15–29.

Feachem, R.G., Bradley, D.J., Garelick, H. and Mara, D.D. (1983). *Sanitation and Disease: Health Aspects of Excreta and Wastewater Management* (London: John Wiley).

IRC (1991). *Just Stir Gently; the Way to Mix Hygiene Education with Water Supply and Sanitation*, IRC Technical Paper No. 29 (The Hague: IRC International Water and Sanitation Centre).

Kalbermatten, J.D., Julius, D.S. and Gunnerson, C.G. (1982). *Appropriate Sanitation Alternatives: A Technical and Economic Appraisal* (Baltimore: Johns Hopkins University Press).

Mara, D.D. and Feachem, R.G. (1980). Technical and public health aspects of sanitation programme planning, *Journal of Tropical Medicine and Hygiene*, **83**, 229–240.

Pacey, A. (1978). *Sanitation in Developing Countries* (London: John Wiley).

Pacey, A. (1980). *Rural Sanitation: Planning and Appraisal* (London: Intermediate Technology Publications).

Pasteur, D. (1979). The Ibadan comfort stations programme: a case-study of the community development approach to environmental health improvement, *Journal of Administration Overseas*, **18**, 46–58.

Perrett, H. (1983). *Planning of Communication Support (Information, Motivation and Education) in Sanitation Programmes*, TAG Technical Note No. 2 (Washington DC: World Bank).

Perrett, H. (1984). *Monitoring and Evaluation of Communication Support Activities in Low-Cost Sanitation Projects*, TAG Technical Note No. 11 (Washington DC: World Bank).

Simpson-Hébert, M. (1983). *Methods for Gathering Socio-cultural Data for Water Supply and Sanitation Projects*, TAG Technical Note No. 1 (Washington DC: World Bank).

Squire, L. and Van der Tak, H. (1975). *Economic Analysis of Projects* (Baltimore: Johns Hopkins University Press).

Wagner, E.G. and Lanoix, J.P. (1958). *Excreta Disposal for Rural Areas and Small Communities* (Geneva: World Health Organization).

10

Waste Water Treatment

10.1 INTRODUCTION

Many forms of treatment are available for either water-borne wastes (sewage) or carted wastes (nightsoil). Some of these have been dealt with in Chapter 8 — for instance, septic tanks and aqua-privies are both forms of treatment, as are composting toilets and biogas plants. An excellent summary of tropical waste water treatment can be found in Mara (1977a) while full details of the design of many treatment systems are contained in Mara (1976).

In this chapter we shall comment only on those systems not adequately covered in temperate climate texts and which are especially suitable for hot climates: waste stabilization ponds, aerated lagoons, and oxidation ditches. Some of the systems used in temperate countries are not suitable for hot climates. Trickling filters, for example, often become an intolerable nuisance due to flies, midges, and odours in hot climates. Besides, they involve moving parts which can break down and are often hard to replace in developing countries, and require periodic cleaning if they are not to become clogged.

10.2 WASTE CHARACTERISTICS

Domestic wastes needing treatment may be either sewage (comprising faeces, urine, and sullage) or nightsoil (comprising only faeces and urine plus small volumes of water if it is used for anal cleansing and pour-flushing). Sewage flows in pipes called sewers and may come from a water-borne sewerage system or from septic tank or aqua-privy effluent. The strength of sewage is usually expressed in terms of its biochemical oxygen demand (BOD), which is defined in Section 3.2.

Raw sewage strengths may be categorized as follows (Mara, 1977a):

Strength	BOD (mg/l)
Weak	200 or less
Medium	350
Strong	500
Very strong	750 or more

BOD values of 400–800 mg/l are common in cities and towns in developing countries. In such areas, raw sewage contains approximately 40 g of BOD/person.day. This BOD comes not only from the faeces and urine but also from the sullage. Thus, if a community is using 100 l/person.day of water, its sewage contains $(40 \times 10^3)/100$ = 400 mg/l of BOD. If the sewage has passed through a septic tank or aqua-privy it is known as settled sewage and it has lost approximately half of its BOD.

Nightsoil, not diluted with sullage, is clearly of lower volume and higher strength. Each person may contribute 1–2 l/day and 30 g of BOD/day. Thus the BOD of nightsoil may be as high as 30 000 mg/l.

10.3 WASTE STABILIZATION PONDS

Ponds are undoubtedly the most widely applicable and advantageous method of waste treatment in hot climates. They consist of a series of shallow lakes through which the sewage flows. Treatment occurs through natural physical, chemical, and biological processes and no machinery or energy input (except the sun) is required. Ponds are the cheapest and simplest of all treatment technologies and are capable of providing a *very high-quality effluent*. In particular, they can reduce levels of pathogenic micro-organisms well below those obtained by other types of treatment; the effluent from a properly designed series of ponds may contain less than 100 faecal coliforms/100 ml. An effluent of this quality may be used for unrestricted irrigation (WHO, 1989).

Ponds are very easy to maintain and require no routine operation. They are able to absorb both hydraulic and organic shock loads and can treat a wide variety of domestic and industrial wastes. They are flexible systems which can be extended to add new capacity if required. Since no imported machinery is required, there is a saving in foreign exchange and problems of obtaining spare parts are avoided.

The greatest disadvantage of ponds is that they take up a lot of space. In a hot and sunny area one might roughly estimate that a total area of 0.3–0.4 ha/1000 persons is required. Thus, a town of 50 000 would require a pond area of 15–20 ha. Great savings in space can be achieved by incorporating anaerobic pre-treatment, but the area required is still substantial when compared with treatment works of the types conventionally used in temperate climates. Sewage treatment works usually have to be on flat land of the sort suitable for building or agriculture, and this may have a high opportunity cost. In areas where land is scarce or very expensive (or both), ponds may have to be rejected. However, it should be borne in mind that often one of the best investments that a municipality can make is to buy land for ponds on the outskirts of the urban area. If the town grows and comes so close to the ponds that they create a nuisance, they may be filled in and sold as building land at a substantial profit and new ponds constructed further out.

Three main types of pond are found in practice and are joined in series. Figure 10.1 shows some typical pond layouts incorporating anaerobic ponds, facultative ponds, and maturation ponds. A fourth type of pond, the high-rate pond, is mentioned below. An aerated lagoon, discussed later in this chapter, is also shown in Figure 10.1.

Anaerobic ponds

Anaerobic ponds are essentially open septic tanks used to provide pre-treatment of large volumes of strong wastes. Anaerobic digestion and settlement take place and a thick scum usually develops on the surface. Retention times are typically in the range 1–4 days and a depth of 2–4 m is preferred. To avoid odour release the volumetric loading (see below) should not exceed 400 g/m^3.day of BOD and the concentration of sulphate ion in the raw waste should not be greater than 100 mg/l (expressed as SO_4^{2-}). At temperatures above 20°C, BOD removed may be estimated as follows:

1   day retention: at least 50% BOD removed

2.5   day retention: at least 60% BOD removed

5   day retention: at least 70% BOD removed

For temperatures in the range of 15–20°C these values should be reduced by about one-quarter (Mara, 1977b).

Anaerobic ponds permit savings of land when large pond systems are being designed. They will reduce the BOD of a strong sewage from, say, 600 mg/l to 240 mg/l and thus greatly reduce the area of the facultative pond required. Wastes having passed through a septic tank or aqua-privy have already undergone the treatment provided in an anaerobic pond and so should flow straight to a facultative pond. Anaerobic ponds, like septic tanks, will accumulate sludge at about 0.03–0.04 m^3/person.year and will require desludging every 3–5 years.

It is recommended that anaerobic ponds be designed to have a volumetric loading rate of 250 g/m^3.day of BOD (Mara, 1977b). The volumetric loading rate λ_v is the daily amount of BOD entering the pond, divided by the pond volume V. If the pond receives Q m^3/day

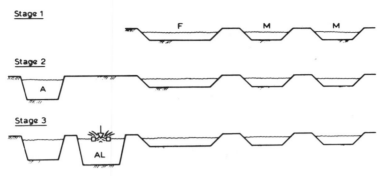

Stage 1

Stage 2

Stage 3

Figure 10.1 Stages in the development of a waste stabilization pond-aerated lagoon system. F—facultative pond; M—maturation pond; A—anaerobic pond; AL—aerated lagoon. At stage 3 additional maturation ponds will probably be necessary. In some cases septic tanks may replace anaerobic lagoons (usually for populations below 10 000)
Source: From Mara (1977a)

of sewage whose BOD is L_i mg/l, then the daily load of BOD is L_iQ g/day. Therefore:

$$\lambda_v = L_iQ/V = 250 \text{ g/m}^3.\text{day}$$

Hence, if the volume and strength of influent sewage are known, the pond volume required can be calculated, and selecting a convenient depth in the range 2–4 m, one can derive the area.

Since the retention time in days t^* is equal to V/Q,

$$t^* = L_i/\lambda_v = L_i/250$$

Therefore the BOD of the influent sewage in mg/l, divided by 250, gives the required retention time. Thus, a sewage of BOD 500 mg/l requires two days, while a very strong waste of 1000 mg/l needs four days.

If a retention time of less than one day is indicated, in other words if the influent has a BOD of less than 250 mg/l, an anaerobic pond should not be used and wastes should pass directly to a facultative pond.

Facultative ponds

The facultative pond is usually the largest pond in the system, and in the absence of pre-treatment in anaerobic ponds, it is the pond into which wastes will flow first (Figure 10.1). In the upper layers of the pond, oxidation of organic matter takes place with oxygen provided by photosynthesizing algae. A symbiotic relationship between the algal and bacterial communities is built up and is illustrated in Figure 10.2. The principal mechanisms for BOD removal are illustrated in Figure 10.3. Sludge accumulates and digests anaerobically at the base of the pond so that desludging is required only every 10–20 years.

To design a pond it is necessary to determine its depth and area. Depth is chosen to be 1.2 m, a compromise between excessive anaerobicity in deeper ponds and the risk of emergent vegetation in

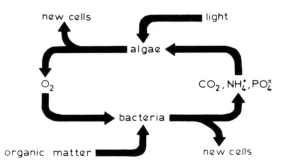

Figure 10.2 Symbiosis of algae and bacteria
Source: From Mara (1977a)

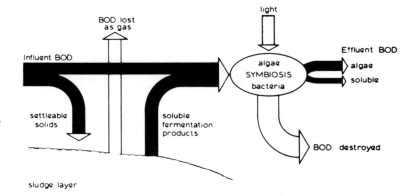

Figure 10.3 Pathways of BOD removal in facultative stabilization ponds
Source: From Mara (1977a)

shallow ponds. The area can be calculated from a number of formulae described by Mara (1976). The following procedure is suitable in most cases and is based on the surface BOD loading rate λ_s (kg/ha.day). The surface and not the volumetric rate is used here, because the capacity of a facultative pond is determined by the amount of sunlight it receives, which in turn determines the algal productivity and thus controls the cycle shown in Figure 10.2. A practical design equation based on considerable experience is the following:

$$\lambda_s = 20\,T - 60$$

where T is the mean monthly ambient air temperature of the coldest month in °C. Hence the design equation for the pond area can be obtained:

$$A = QL_i/(2T - 6)$$

where A= pond area (m^2);

       Q= flow of sewage (m^3/day);

       L_i= BOD of influent (mg/l).

If the influent BOD is not known from past experience it may be calculated from the per capita water use and the BOD contribution per capita (typically 40 g/person.day of BOD).

Example: suppose a facultative pond is required for 50 000 people using 100 l of water per capita per day in northern Nigeria.

$$L_i = \frac{40 \times 10^3}{100} = 400 \text{ mg/l}$$

$$T = 21°C \text{ (from meteorological records)}$$

$$Q = 50\ 000 \times 100 \times 10^{-3} \text{ m}^3/\text{day}$$

$$\therefore \quad A = \frac{50\ 000 \times 100 \times 10^{-3} \times 400}{(2 \times 21) - 6}$$

$$= 56\ 000 \text{ m}^2$$

$$= 5.6 \text{ hectares}$$

Nightsoil may also be treated in a facultative pond by dumping it on a concrete ramp and sluicing it down into the pond with a jet of water. Water lost by evaporation and seepage must be replaced to maintain a depth of 1.2 m. Pond area may be calculated by the same method. Taking nightsoil to contain 30 g of BOD/person.day:

$$A = \frac{30P}{(2T - 6)}$$

A = pond area (m^2);

P = population;

T = mean ambient air temperature of coldest month($°C$).

Example: for a nightsoil cartage system for the same population of 50 000 in northern Nigeria:

$$A = \frac{30 \times 50\ 000}{(2 \times 21) - 6}$$

$$= 42\ 000 \text{ m}^2$$

$$= 4.2 \text{ hectares}$$

The BOD of the pond effluent can be calculated, if necessary, from another empirical formula:

$$\lambda_r = 0.725\lambda_s + 10.75$$

where λ_r = BOD removal (kg/ha.day)

Maturation ponds

A facultative pond treating sewage or nightsoil must *always* be followed by two or more maturation ponds (Figure 10.1). Maturation ponds are wholly aerobic and are responsible for the final improvement in chemical quality (BOD removal) and for most of the reduction in the numbers of faecal bacteria and viruses. Although sophisticated design methods are available, a good rule of thumb is to provide three maturation ponds, each with a retention time of five days and a depth of 1–1.5 m. Four maturation ponds, each with a four-day retention time and the same depth (thus having a similar total area) provide a greater degree of microbiological purification. In a warm climate, each pond with a five-day retention time removes at least 95% of the faecal coliforms entering it. Maturation ponds provide excellent environments for fish-breeding (see Section 14.3).

High-rate ponds

These are as yet an experimental system for the rapid conversion of organic wastes into algae in shallow ponds. They are usually linked to some device for algal harvesting, which is a complex technology. They are not generally applicable to simple waste treatment works in developing countries.

Bacterial reduction

The reduction of faecal bacteria in a pond (anaerobic, facultative, or maturation) has been found to follow an equation of the form

$$N_e = \frac{N_i}{1 + Kt^*}$$

where N_e = number of bacteria/100 ml of effluent;

N_i = number of bacteria/100 ml of influent;

K = a constant for a given species, temperature etc.;

t^* = retention time in days.

For faecal coliforms, a reasonable design value of N_i is 10^8 FC/100 ml; this is slightly higher than average values normally found in practice.

The 'rate constant' K represents the rate at which the bacteria die off, and is extremely temperature-sensitive. For faecal coliforms, it may be estimated from the equation

$$K = 2.6(1.19)^{T-20}$$

where T is the temperature in °C. In anaerobic ponds, a lower rate constant of approximately $K/2$ is appropriate.

For a system with an anaerobic pond, a facultative pond and n maturation ponds, the equation becomes:

$$N_e = \frac{N_i}{(1 + (K/2)t_a^*)(1 + Kt_f^*)(1 + Kt_{m_1}^*)\ldots(1 + Kt_{m_n}^*)}$$

where t_a^*, t_f^*, etc. are the retention times of successive ponds in the system.

In practice, the retention times of the anaerobic and facultative ponds will be determined by other criteria, described above. If the value of N_e is unacceptable, then the size or (preferably) the number of the maturation ponds must be increased until the effluent quality requirement is satisfied.

Location

The location of a system of ponds is largely determined by local topography and the availability of a suitable site. Houses and schools have been successfully located within 100 m of well-designed and well-operated facultative ponds, but unacceptable odour and insect nuisances depend on the standard of maintenance attained, and may extend further than this, particularly from anaerobic ponds. In India, facultative ponds are located at least 450 m from large residential areas, and in the USA a minimum distance of 0.5–1 km is advocated.

Design features

One facultative pond followed by two or three maturation ponds is a normal arrangement. The inlet pipe to the facultative pond should discharge below the water level, to prevent the formation of excessive scum by entrainment of air. It should be carried on pillars to at least 10 m from the pond edge, to prevent it from becoming blocked by the sludge which will accumulate below it. Typical interpond connections are shown in Figure 10.4. Ponds should be rectangular with a length to breadth ratio of 2–3 to 1. If excessive seepage losses (more than 10%) from the pond base are anticipated, the base should be lined with puddled clay, polythene sheeting, bitumen, or other appropriate material. It should be remembered that permeable bases in facultative ponds will soon seal up as sludge accumulates, although this will not occur to such an extent in maturation ponds.

Embankments are usually made with slopes of 1 in 2–3 and should be protected at water level from erosion and vegetation growth. Concrete slabs, stabilized soil, or stone rip-rap are appropriate (Figure 10.5). Surface runoff must be prevented from flowing into the ponds during heavy rainfall and it may be necessary to provide drainage ditches around the ponds. In some arid areas, wind-blown sand tends to fill the ponds. This must be prevented

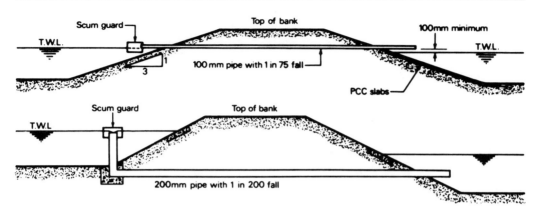

Figure 10.4   Two examples of simple interpond connections (PCC: Precast concrete)
Source: From Mara (1977b)

by building wind breaks. A fence or hedge is required, with clear warning signs, to prevent access to the ponds by people and animals.

Pre-treatment

Screening is advisable for sewage entering a waste stabilization pond system. Screens may be raked manually. Grit removal followed by comminution (a shredding process) is an alternative although less appropriate due to the mechanical sophistication of the comminuter and the need for a power source. Where the grit content is high, grit-removal channels should be provided.

Pond maintenance

Maintenance consists mainly of controlling vegetation growth, maintaining a fence around the ponds, and removing any scum mats which may form. Vegetation tends to grow down the banks and into the pond edges (Figure 10.5). It is essential to keep the banks free from vegetation and the surrounding area tidy. If ponds become overgrown, not only will their performance be hindered, but breeding sites for mosquitoes and snails will be formed and these may promote the transmission of filariasis, schistosomiasis, and possibly other infections — in addition to constituting a nuisance. For a small pond system, one or two labourers, *under good supervision*, are perfectly adequate to carry out maintenance tasks.

10.4 AERATED LAGOONS

If a facultative pond is too small, or if due to toxic substances or lack of sunlight the algae are not adequately photosynthesizing,

a

Figure 10.5 (a) A well-constructed and well-maintained pond bank, also showing inlet arrangement suitable for a maturation pond. (b) A very poorly constructed and badly maintained stabilization pond with heavily overgrown banks providing breeding habitats for snails and mosquitoes
(Photos: D Mara)

b

the biochemical oxygen demand will exceed the oxygen supply and the pond will turn anaerobic. This may be prevented by providing mechanical aeration; such a system is known as an aerated lagoon. Oxygen is provided by motor-driven surface aerators, and the lagoon develops a flocculated suspension of bacterial cells.

In aerated lagoons, the organic solids in sewage are converted into bacterial cells and these cells form a sludge. This sludge must be removed before the effluent is discharged or re-used, and therefore aerated lagoons are generally followed by maturation ponds. Aerated lagoons are analogous to activated sludge plants without recycling of the sludge. Four days is a typical retention time in aerated lagoons and will remove 85–90% of BOD. Bacterial reduction is poor but will be made excellent by the provision of sufficient maturation ponds.

Preliminary treatment by screening, or by grit removal and comminution is required. Depths of 3–4 m are preferred with bank slopes of 1 in 2. Banks and the lagoon bottom must be protected from erosion caused by the turbulence set up by the aerators.

Aerated lagoons incorporate sophisticated mechanical and electrical equipment and therefore lack the benefits of simplicity possessed by waste stabilization ponds. They are most appropriate for extending the capacity of an existing pond treatment system as indicated in Figure 10.1. Design details may be found in Mara (1977b).

10.5 OXIDATION DITCHES

Oxidation ditches are similar in principle to aerated lagoons. However, the layout is quite different and most of the sludge is recirculated. A typical ditch plan is shown in Figure 10.6. Wastes are circulated round an oval channel 1.5–2 m deep at a velocity of about 0.3–0.4 m/s. The velocity and the aeration are provided by rotating cylindrical 'brushes' which drive the water forward while, at the same time, providing intense turbulence. Effluent from the ditch is settled in a secondary sedimentation tank and > 95% of the sludge from this tank is returned to the ditch.

Recirculating the sludge back into the ditch produces a much richer concentration of bacterial flocs than would be produced in an aerated lagoon. This facilitates shorter retention times (1–3 days) and causes the sludge to be under aeration for much longer periods (20–30 days). This produces a sludge excess which is highly mineralized and can be placed directly on sludge-drying beds without further digestion. BOD reduction is typically good and an effluent BOD of less than 15 mg/l can usually be produced. Bacterial removal is poor and the effluent must be discharged to a large body of receiving water or further treated in maturation ponds before use in irrigation.

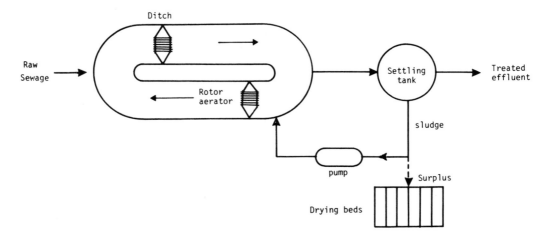

Figure 10.6   Typical oxidation ditch installation
Source: From Mara (1976)

Details of oxidation ditch design are given by Mara (1976). They are compact systems and produce low sludge volumes, but they require complex machinery and substantial amounts of electric power, and present operation and maintenance problems equivalent to those of a conventional activated sludge or trickling-filter plant.

10.6 PATHOGEN REMOVAL

Most engineers have been educated to design sewage treatment plants on the basis of physical and chemical properties, such as the removal of suspended solids and BOD. Where effluents are to be re-used for irrigation, or discharged into rivers or lakes which local people use for drinking water or recreation, it is imperative to design primarily on the basis of pathogen removal. Raw sewage may contain considerable concentrations of the pathogens associated with all infections in Categories I–V in Table 1.3. The removal or significant reduction of these pathogens is a stringent requirement and, in general, a treatment plant with a good pathogen-removal performance also has a good physico-chemical performance, although the converse is not true.

The subject of pathogen removal in waste treatment processes is too complex to be discussed fully here. A detailed account is given by Feachem *et al.* (1983). The most important fact associated with sewage treatment is that conventional treatment plants (primary sedimentation followed by trickling filters or activated sludge followed by secondary sedimentation) remove only between 90 and 99% (1–2 $\log_{10}$ units) of faecal viruses and bacteria. This is *a very poor level of removal* and turns an influent containing 10^7

faecal coliforms/100 ml into an effluent containing 10^5–10^6 faecal coliforms/100 ml.

By contrast, a well-designed system of waste stabilization ponds achieves high orders of pathogen removal of 99.99% (4 $\log_{10}$ units) or better. A series of five ponds in northeast Brazil, with a total retention time of twenty-nine days and an average temperature of 26°C, reduced faecal coliforms by 99.999 96% (6.4 $\log_{10}$ units) and produced an effluent with only seventeen faecal coliforms/100 ml (Mara and Silva, 1979). This bacteriological quality is better than found in most surface waters (Table 3.1) and the effluent could be used with confidence for irrigation.

As discussed in Chapter 3, the problem of pathogen removal is solved not by designing conventional plants and chlorinating the effluent, but only by installing a plant that has inherently good pathogen-removal performance or by discharging the effluent to a large body of water not used for domestic or recreational purposes.

10.7 SEWAGE WORKERS' HEALTH

There have been several investigations into the health of sewage plant workers. Results are often contradictory, but some studies have shown sewage workers to have an increased risk of contracting diarrhoeal diseases and intestinal worms (Table 1.3, Categories I, II, and III), as well as leptospirosis. It should therefore be assumed that sewage workers are at risk and steps should be taken to protect them. These steps involve the good maintenance of the plant, attention to general tidiness and hygiene, immunization, the provision of overalls, boots, and gloves, the provision of washing facilities and health education for the workers. A careful watch should also be kept on the health of the workers' families. Regular treatment for intestinal worms should be the minimum provision; the drugs now available are cheap, effective, and safe.

10.8 REFERENCES AND FURTHER READING

Arthur, J.P. (1983). *Notes on the Design and Operation of Waste Stabilization Ponds in Warm Climates of Developing Countries*. World Bank Technical Paper No. 7 (Washington DC: World Bank).

Clark, C.S., Cleary, E.J., Schiff, G.M., Linnemann, C.C., Phair, J.P. and Briggs, T.M. (1976). Disease risks of occupational exposure to sewage, *Journal of the Environmental Engineering Division, ASCE*, **102**, 375–388.

Feachem, R.G., Bradley, D.J., Garelick, H. and Mara, D.D. (1983). *Sanitation and Disease: Health Aspects of Excreta and Wastewater Management* (London: John Wiley).

Mara, D.D. (1976). *Sewage Treatment in Hot Climates* (London: John Wiley).

Mara, D.D. (1977a). Waste water treatment in hot climates, in *Water, Wastes and Health in Hot Climates*, Feachem, R., McGarry, M. and Mara, D. (eds) (London: John Wiley), pp. 264–283.

Mara, D.D. (1977b). *Sewage Treatment in Hot Countries*, Overseas Building
    Note No. 174 (London: Building Research Establishment).

Mara, D.D. and Silva, S.A. (1979). Sewage treatment in waste stabilization
    ponds: recent research in northeast Brazil, *Progress in Water Technology*,
    11, 341–344.

USEPA (1983). *Design Manual; Municipal Wastewater Stabilization Ponds*.
    Document EPA-625/1-83-015. (Washington DC: US Environmental
    Protection Agency.

WHO (1987). *Wastewater Stabilization Ponds; Principles of Planning &
    Practice*, WHO EMRO Technical Publication No. 10 (Alexandria, Egypt:
    World Health Organization).

WHO (1989). *Health Guidelines for the Use of Wastewater in Agriculture
    and Aquaculture*, WHO Technical Report Series No. 778 (Geneva: World
    Health Organization).

11

Surface Water Drainage

11.1 INTRODUCTION

In the slums and shanty towns of the Third World, many residents feel that they need drainage more urgently than water supply or latrines. This is partly because the only land which they can afford, or on which the owners will allow them to stay as squatters, is land which is unsuitable for building. It may be on steep hillsides subject to erosion and landslides, as in parts of Rio de Janeiro or Luanda. Soil eroded from a hillside is usually deposited at the foot of the slope; soil eroded in a single rainstorm has been known to bury houses completely in this way. More often, the land is low-lying, marshy and subject to frequent flooding, often with faecally polluted floodwater. In cities such as Bangkok, Calcutta, Dar es Salaam, Lagos and Recife, many neighbourhoods are flooded several times a year, and people have to cope with water or other people's sewage inside their dwellings. Sometimes the residents build their houses on stilts and connect them by elevated pathways, although these are often rickety and unsafe.

Historically, most of the major cities of the developing world arose along the coast as ports, often on the estuaries of rivers which served as commercial arteries for the transport of goods to and from the hinterland. The flat estuarine terrain and soft, often impermeable alluvial soils make drainage difficult, and the coastal regions of the world are where the highest average rainfall is found.

Even in the arid areas where average rainfall is low, tropical rainfall — when it comes — is more intense than in temperate climates, and the lack of vegetation and of adequate drainage means that torrents of water can arise in minutes, causing damage to homes and property which may take years to repair. Not only is rainwater a problem. Leaking water mains, waste water from washing and bathing, and the sewage from overflowing septic tanks and blocked sewers constitutes a health hazard, damages buildings and can cause

Table 11.1 The main constituents of urban surface water, and the control strategies appropriate to each

Surface water constituent	Characteristics and problems	Physical control strategies
Stormwater	high flow, short duration, seasonal	drains, fill, and pumping
External floodwaters	high flow, short duration, seasonal	dykes, fill, polders, training walls
Marsh water	long-term ponding	dykes, fill, pumps, polders, drains
Sullage	low flow, steady, not seasonal, some public health risk	sullage ditches, discharge to storm drains, soakaways
Toilet wastes	low flow, steady, not seasonal, high public health risk	sullage separation and low-cost sanitation; septic tanks and soakaways; sanitary sewers and treatment; combined sewers and treatment of dry weather flow

Source: From a report by P. J. Kolsky

flooding if an adequate drainage system does not exist. Table 11.1 lists the main constituents of urban surface water, each with its own problems and drainage requirements.

Less dramatic than drowning in floods or burial beneath landslides and collapsing homes, but no less important in the residents' lives, is the toll of disease, disability and death caused by standing water in a poor community. First in public health importance are the many faecal–oral infections (Category 1, Table 1.2) which can be transmitted by surface water contaminated with excreta from blocked sewers and overflowing septic tanks, and often from defecation in the open by livestock and by people who have no toilet. Children are particularly exposed to infection when playing or bathing in the water. Improved water supply and excreta disposal are almost impossible to install in areas subject to frequent flooding. Poorly drained urban areas also present ample opportunities for transmission of schistosomiasis, in those countries where it is endemic (see Chapter 17).

Poor drainage also favours the breeding of mosquitoes, and hence the transmission of mosquito-borne infections (Chapter 15). Transmission can be especially intense in urban areas where there are relatively few animals to divert the vector species of mosquitoes from human blood meals. Bancroftian filariasis is particularly closely associated with poor urban drainage. In some Indian cities, more than half the *Culex quinquefasciatus* mosquitoes which transmit the disease breed in the drains — supposedly built to control mosquitoes.

Anopheline mosquitoes do not usually breed in heavily polluted water, but can multiply in swamps, pools, puddles, and also in streams and stormwater canals in which there is standing water. Anopheline mosquitoes breeding in poorly drained low-income areas can transmit malaria to adjacent wealthy parts of town. A particular danger of malaria transmission in a city is the large amount of population movement to and from it, which increases the risk of the importation of new and possibly drug-resistant strains of the disease.

Drainage construction is an effective mosquito control measure. It requires a one-time capital investment, followed by recurrent costs for maintenance which may be minimal if a good level of community participation can be obtained. In many cases, the initial investment costs less than one year's supply of insecticide. The difference will become more pronounced in the future, with the continued spread of insecticide resistance, and the consequent need for more expensive compounds. In fact, drainage construction can help to slow down the development of insecticide resistance by reducing the amounts of insecticide required. The use of insecticides for public health purposes usually carries little environmental risk, but drainage construction is, by definition, an environmental improvement. Compared to chemical control programmes, environmental modifications such as drainage systems are much less affected during periods of political, social or economic instability.

Poor drainage is not only an urban problem. There is increasing awareness of the need to include drainage improvements with other environmental sanitation measures in rural communities also (Muir, 1986).

11.2 TOWN PLANNING IMPLICATIONS

The urban poor often build on land with drainage problems, but good urban physical planning can help to avoid making those problems worse.

One of the simplest planning measures is to set out regular plots before house building starts in an area, leaving space for well-aligned roads. Adequate road width and alignment makes it much easier to build drains when they are needed later. Site-and-service schemes are expensive and take a long time to plan and implement, but such a 'site only' scheme should be within the means of any municipality. Once the overall layout of a neighbourhood has been planned, residents (or future residents) can be shown how to set out individual rectangular plots with nothing more sophisticated than a tape measure, or even a piece of string with knots at regular intervals.

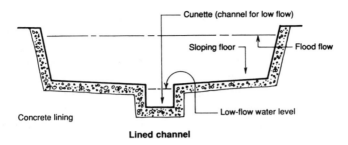

Lined channel

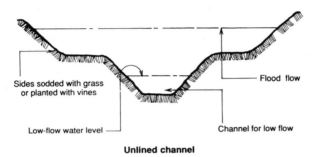

Unlined channel

Figure 11.1
Composite channel
sections designed to
take sullage or septic
tank effluent in the
central 'cunette', and
so ensure self-cleaning
flow in dry weather
and prevent mosquito
breeding

A 3–4–5 triangle can be used to mark the first right angle. Some degree of discipline over house building is necessary, to ensure that plot boundaries are observed, to prevent houses from obstructing existing drainage paths, or from occupying land needed for future drainage works. The residents themselves are in the best position to enforce this.

The development of residential areas can increase drainage problems in several ways. As vegetation is removed, the capacity of the ground to retain water and resist erosion is reduced. The increasing area covered with roofs and road surfaces diminishes the area of ground into which water can infiltrate, leaving a greater volume of water to be removed by drainage. Low-lying areas subject to flooding play a role in storing the water from sudden rainstorms until it can drain away gradually; when these are filled in for housing, this may cause flooding in other areas.

Roads must be built above the flood level, and the embankments this requires can obstruct natural lines of drainage, or can channel water alongside them causing erosion. In some cases, as in parts of Bangkok, roads have been built by filling in existing channels, causing serious flooding. Where the natural drainage channels are not filled in or obstructed by buildings, they often become blocked by domestic refuse.

On the other hand, drainage improvements in one area are closely interrelated with drainage problems elsewhere, and are best planned in the context of the city as a whole, or at least of a whole catchment area. Better drainage in one neighbourhood means that surface water flows away faster, imposing a greater burden on the capacity of the system downstream. At the same time, drainage improvements at a local level may be of little use if water still backs up because this downstream capacity is insufficient. This has been a serious problem in Jakarta, where improved local drains have often been submerged by water held back by constrictions in the city's major canals.

Of course, it is possible for a community to make local improvements, even without the support of the city planning authorities, but at least some consideration should be given to the body of water into which a new drainage system will discharge. Whether this is a main sewer, river, lake or sea, the maximum level to which it floods will normally set the minimum level for the drainage system. The discharge of drainage water also affects the quality of the 'receiving water' into which it flows, especially when sewage or septic tank effluent is released into the drains. In Bangalore, for example, the discharge of sewage into several dams in the city led to intense breeding of *Culex* mosquitoes until measures were taken to breach or bypass them.

11.3 TECHNICAL ASPECTS

The technical considerations affecting the design and construction of the drainage system of a tropical town are somewhat different from those faced in an industrialized country. Conditions are especially different where low-income communities are concerned. The main differences are listed below.

Resources   Whereas the cities of Europe and North America took many years and considerable funds to build up their drainage systems, the conurbations of the tropics have neither the time nor the money. Not only are they in poor countries; the municipalities are poorer still, so that even were they to receive aid funding for drainage construction, they would have scant resources with which to carry out maintenance. Labour, on the other hand, is relatively cheap, so that labour-intensive construction and maintenance techniques are more appropriate.

Objectives   The rich nations can afford to pay for, and expect, drainage of a very high standard. Flooding of the streets, even on one day in a year, is unacceptable. In developing countries, on the other hand, significant improvements can be obtained with quite modest objectives. In order to control mosquito breeding, for example, it is sufficient to remove surface water within a few days and not immediately. A few inches of flooding several times each year may be a great improvement on waist-deep water for weeks on end.

Hydrology   Tropical rainfall patterns are different from those in the temperate zones. In many tropical climates, it is not uncommon for half the average annual precipitation to fall in a single rainstorm. The best option may be to allow for flooding in such a storm. Otherwise, very large drainage channels may be needed, which will be almost empty for most of the time. They will then need thoughtful design, to prevent mosquito breeding and ensure self-cleaning flows during dry weather (Figure 11.1).

Channels   The need to avoid overloading of closed sewers, a major design consideration in western cities, is not relevant to a drainage network of open channels. This means that a drainage system can be designed to overflow from time to time; even a flooded system removes water faster than no system at all. Channels also have considerable storage capacity which can attenuate peak flows.

Sediment   Rainfall intensities are generally higher in tropical cities, but the paved area of low-income urban settlements is often quite low in relation to population density. Both these factors contribute to a much higher sediment load than is found in the runoff of temperate cities. This can cause rapid blocking of even large open drains unless special precautions are taken. In such conditions, a conventional silt trap at the entrance to a culvert would be a waste of money, as it could fill with sand in a matter of hours.

Maintenance   Drainage needs more maintenance in the tropics. The rapid growth of tropical vegetation can block a drain quickly if it is not regularly cleared. When solid waste collection services are deficient, garbage will find its way into the nearest available ditch, and can accumulate to block it within days if it is not removed (Figure 11.2).

Organization   On the other hand, poor urban communities in the Third World generally have a far greater degree of coherence and organization at the neighbourhood level than is to be found in the industrialized countries. This creates a potential for community

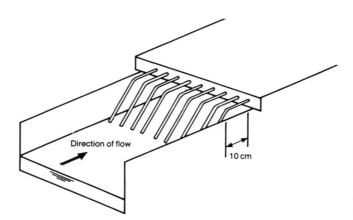

Direction of flow

10 cm

Figure 11.2 A grating designed for easy cleaning with a rake. Without such protection, culverts and other covered sections of channel quickly become blocked

participation which could be mobilized for the construction and maintenance of drainage works. In some cases, it has been very effective.

One promising drainage arrangement is to use the road itself as a wide, shallow drainage channel, .or to lay channels beneath roads or access ways (Figure 11.3). Prefabricated channel linings can be made in a central workshop, but laid by local residents under supervision (Figure 11.4).

11.4 INSTITUTIONAL ASPECTS

Drainage improvements are not only a job for a drainage engineer. They involve several professions and need the cooperation of several sectors if they are to succeed. Drainage is of concern to town planners, and if some houses have to be relocated to make room for new drains, architects and builders may also be involved. Drains are usually built beside roads, and the roads department will have an interest because good drainage is essential to protect the road surface.

Maintenance of the drainage system depends on an efficient solid waste collection service, as without one the drains will soon fill with rubbish. Moreover, the street cleaning and solid waste collection service will often be the most suitable municipal department to clean the drains regularly, as it will have the necessary vehicles to remove

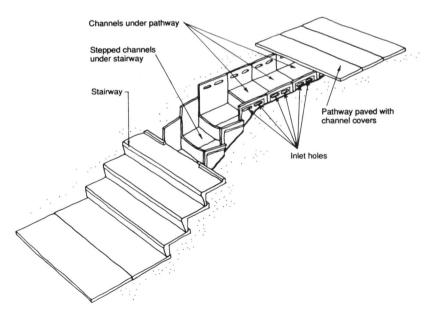

Figure 11.3   Combined drainage channel and access way as used in Salvador, Brazil. The precast modular sections composing the system are each designed to weigh less than 50 kg, so that they can be carried to site by two people

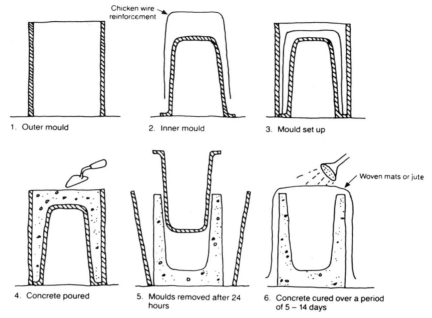

(*a*) **The Bangkok drain element**

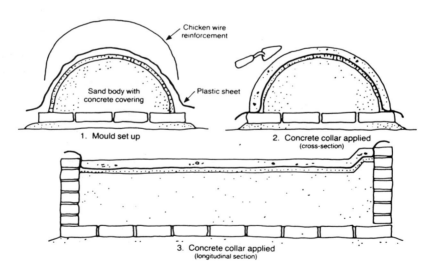

(*b*) **The Roorkee drain element**

Figure 11.4   Some methods for the production of prefabricated drainage elements

the solid materials such as silt, vegetation and refuse which will accumulate in them. The health department will be concerned to ensure that the cleaning is done well and regularly, and that the drains are not built in such a way as to make this difficult or to

promote disease transmission. This in turn involves several special-
ities, including medical entomology.

The community has a key role to play (Figure 11.5). Their
cooperation is needed in obtaining the necessary land; some people
may have to agree to relocate their houses to make room for the
new drains. Whether or not the community takes responsibility for
maintaining the system, a responsible attitude on their part will be a
great help towards its upkeep, reducing the amount of rubbish thrown
into the drains, and damage done to them by vehicles, building work
or vandalism. A single uncooperative resident who blocks the water
flow, or neglects to clean his section of the drainage line, can harm
the interests of the whole community. Proper drainage therefore calls
for the close cooperation of the community and their leaders, and
also of those who work with the community, such as educators and
health workers.

A cooperative attitude, however, is not enough. Effective collaboration between municipal departments and involving the community
has institutional implications.

At the level of local government, *some* department at least must
have the primary responsibility for urban drainage. In many cities

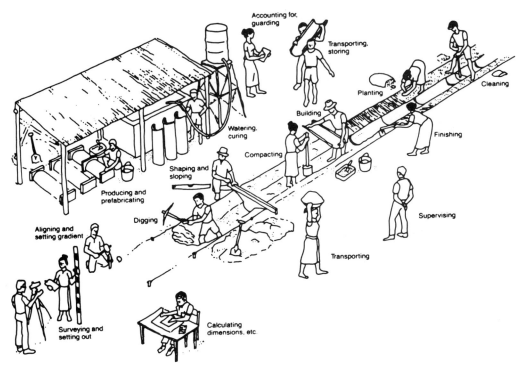

Figure 11.5 Tasks in the construction of a drainage system. Most of these can be carried out by members of the community, if they are given help with the others

there is no clear definition of who is responsible for cleaning and maintaining the natural and man-made drainage system, and in some it is not even clear who is to build it, or which national government department is to finance major drainage works. An example of the absurd situations which can arise in such cases is for one department to clean rubbish from the drains, for the solid waste collection service to refuse to collect it from where it is dumped on the adjoining roads, and for the roads department to sweep it back into the drains again!

Ideally, the regular cleaning of urban drains should be the job of the street cleaning and solid waste collection service. However, other sectors will usually be responsible for drainage construction and repairs, so that several sectors are inevitably involved. Some arrangement for regular liaison meetings should therefore be set up, and a single department should be responsible for convening them. The health department should be represented.

Some institutional arrangements are also needed in the community, to mobilize and coordinate the community's contribution and to ensure that it is not undermined by the antisocial behaviour of a minority. It is best to build on existing community institutions if possible, although these may already be fully occupied with other day-to-day tasks. In many cases, a useful initial step will be to form a drainage committee to organize the community's contribution to planning, implementation and maintenance of drainage improvements.

11.5 REFERENCES AND FURTHER READING

Cairncross, S. (1986). Urban drainage in developing countries, *Parasitology Today*, **2**(7), 200–202.

Cairncross, S. and Ouano, E.A.R. (1991). *Surface Water Drainage for Low-Income Communities* (Geneva: World Health Organization).

Manoharn, S. (1982). Application of ferrocement drainage flume in slum upgrading, *Journal of Ferrocement*, **12**(4), 373–383.

Muir, J. (1986). Rural health in northern Pakistan, *Waterlines*, **5**(2), 10–14.

Watkins, L.H. and Fiddes, D. (1984). *Highway and Urban Hydrology in the Tropics* (London: Pentech Press).

WHO (1987). Application of Environmental Management Measures for Disease Vector Control in Urban Areas. WHO/VBC/SEM/VCT/87.8.2 (Geneva: World Health Organization).

12

Refuse Collection and Disposal

12.1 INTRODUCTION

'Refuse' refers here to domestic and industrial solid wastes. Refuse disposal is more a problem of cities than of rural areas. Tropical cities in developing countries produce less refuse per capita than temperate cities because there is less industrial activity, because the inhabitants have less purchasing power and therefore consume less, and because there is a very high rate of re-use by the poorer sections of the community. Despite this, large volumes of refuse are produced and form an enormous public health and aesthetic problem in most developing country cities. Large unplanned dumps of refuse can be seen in many cities and these provide support for communities of people, goats, rats, and dogs who scavenge on them (Figure 12.1).

12.2 WASTE CHARACTERISTICS

The amount of domestic refuse collected varies widely, depending on many factors such as standard of living and eating habits. In Europe and North America it may amount to as much as 2 kg/person.day, but in tropical cities it usually varies between about 0.3 and 1 kg/person.day.

The composition of the refuse also varies. Table 12.1 summarizes the composition of town refuse in India and Europe. It will be seen that India produces less refuse per capita, and that this refuse is considerably denser than in Europe. It contains less plastics, metals, glass, and paper, which are more prone to be re-used by the urban poor.

In India, larger communities produce a greater proportion of compostable matter, and a lower proportion of ash and fine earth, than smaller communities. A daily collection of 0.7 kg/person of compostable refuse has been recorded for Bombay, but 0.2–0.35 kg/person.day is more common in small towns and villages.

Figure 12.1   Goats scavenging on an urban refuse damp in Nigeria
(Photo: R Feachem)

Much industrial waste is similar in composition to domestic and commercial refuse although some industries produce toxic or inflammable wastes which require special disposal facilities. In Britain 7% of industrial wastes are either toxic, caustic, acidic, or inflammable.

Wastes from agriculturally based industries often pose special problems. Tanneries, abattoirs, and canning factories, for instance, produce large volumes of organic wastes which rapidly become offensive in hot climates if not properly dealt with. Pickford (1977) reports that, in Dubai, the unusable abattoir waste from a sheep is 4.7 kg, from a cow 32 kg and from a camel 90 kg.

12.3 HAZARDS OF REFUSE MISMANAGEMENT

A variety of environmental hazards is associated with the mishandling or mismanagement of refuse. Fly-breeding will always

Table 12.1 Approximate compositions of town refuse in India compared to Europe

Characteristic	India	Europe
Paper (%)	2	27
Plastics (%)	1	3
Metals (%)	0.1	7
Glass (%)	0.2	11
Ash and fine earth (%)	12	16
Rags (%)	3	3
Vegetable matter (%)	75	30
Other (stones, ceramics, etc.) (%)	7	3
weight kg/person.day	0.1–1.0	0.8–2.6
Density kg/l	0.33–0.57	0.13–0.27

Note: These figures are taken mainly from Flintoff (1976). Wide variations exist in practice. Refuse from different towns and from different socio-economic groups in the same town, cannot be assumed to be the same, and its contents should be analysed before decisions about disposal and composting are made.

be encouraged by uncovered piles of rotting refuse and the flies may play a role in the mechanical transmission of faeces and thus of faecal-oral diseases (Table 1.2, Category 1). Piles of refuse will also contain mosquito-breeding sites where pools of rainwater form in cans, car tyres, etc. The mosquito *Aedes aegypti* will breed under these conditions and may transmit dengue, yellow fever, and other arboviral infections (Table 1.2, Category 4). Rats will also breed and live in and around refuse. They may promote or transmit a variety of diseases including plague, leptospirosis, flea-borne typhus, ratbite fever, and salmonellosis.

Badly managed refuse can promote water pollution by rain washing debris out of piles of refuse and into surface water. Ground water pollution may also occur. In addition, piles of refuse present a fire risk, they smell, and they are aesthetically unpleasing in the urban environment.

Where refuse disposal services are lacking, much refuse is deposited in open street drains and urban waterways (Figure 12.2). This causes them to block and can cause flooding. It also creates ideal breeding grounds for mosquitoes of the *Culex pipiens* group, which in some parts of the world are the vectors of Bancroftian filariasis (Chapter 15).

12.4 STORAGE AND COLLECTION

Planning and introducing an organized refuse collection service is a complex business. Proposed methods should if possible be tested at pilot scale before extension by stages to cover a whole city.

In high-income, low-density suburbs each house should have

Figure 12.2   Refuse collection cannot be limited to the emptying of dustbins. In this shanty town in Ecuador, as in many other places, it must be dug out of the drainage ditches before the start of the rainy season if flooding and massive mosquito-breeding are to be avoided
(Photo: WHO)

a covered bin, which is emptied once a week, or at some other appropriate interval, into a truck. In some cities, such as Manila, it has been possible to persuade the public to use two bins, one for 'dry' wastes and one for 'wet', which facilitates re-use and disposal. Similarly in many Chinese cities, organic and inorganic refuse are collected separately.

In high-density suburbs it is not possible to empty individual household bins, so that public refuse bins should be provided for each group of houses. These may be permanent fixtures which are emptied by labourers with spades into the collection truck. Preferably, they are metal bins, which are picked up and emptied into a truck with special lifting tackle, or are completely removed and replaced by new ones. Only a system of this kind guarantees proper emptying.

Any lid or door on the bins which requires effort to open will be left open and fly screens will be broken. Therefore it is virtually impossible to prevent fly-breeding in these communal bins. However, it can be reduced by ensuring they are emptied daily, or at least several times a week.

Obtaining public co-operation in the use of bins can be difficult and requires considerable effort by community workers, but can greatly improve the efficiency of the service and reduce its cost. It may be feasible to employ a full-time worker to look after a public bin in the same way as is necessary for a public latrine. Bins should always be protected from the flooding caused by heavy rain.

A fleet of appropriate vehicles is necessary to collect the contents of the bins, or to pick up the bins themselves. The efficient maintenance and operation of these vehicles is one of the most difficult tasks facing the municipality. Some cities have reported over 50% of their vehicles broken down at any one time. Mobile repair teams, sufficient spare vehicles to allow for breakdowns, simplicity and standardization of vehicle design, and regular replacement of old vehicles are all important considerations.

The relatively high density of refuse found in low-income communities (see Table 12.1) means that compaction equipment is unnecessary on the collection vehicles. Compaction gear is a common source of breakdowns, and increases the weight on the rear axle, leaving less capacity to carry the weight of refuse. Compaction vehicles should only be purchased after careful study. An agricultural tractor and trailer is often more practical, and easier to maintain, than many types of purpose-built truck.

Hand- and animal-drawn vehicles are of course cheaper to buy, and more reliable to operate, and more labour-intensive than motorized ones. Their operation costs may be higher, however. In India, 75–80% of the total cost of a refuse system is in transportation, including the cost of transfer, where necessary, from small to large vehicles. It is clear that the correct design and operation of collection

vehicles and transfer stations are of the utmost importance. It should
be remembered that a municipality cannot expect public cooperation
if it is not itself providing a reliable service.

12.5 TREATMENT AND DISPOSAL

The three main methods of treating and disposing of refuse are:
(i) controlled tipping (with land reclamation), (ii) incineration, and
(iii) composting.

Controlled tipping

Controlled tipping, known as sanitary landfill, may be used both
to dispose of refuse and to reclaim excavations, quarries, gravel
pits, swamps, or other suitable sites. Carefully planned tipping may
also be used for flood control. Substantial experience of controlled
tipping has been gained in the last 100 years. There are now standard
procedures for managing a tipping site, and labour-intensive methods
may be used to reduce the need for heavy machinery.

Many tropical cities are in coastal areas where high ground water
levels rule out the use of deep pits. In these cases, a system of parallel
trenches 1 m or more deep can be used, as shown in Figure 12.3.
Further details of this method are given by Flintoff (1984).

Scavenging is almost unavoidable on a tip, and is more easily
controlled than avoided. It provides employment, generates income,
and may reduce the cost of refuse collection. For example, it is

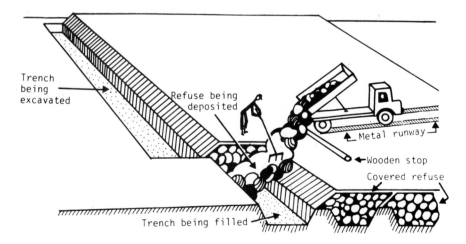

Figure 12.3   The trench method of controlled tipping. The heaps of refuse are levelled and formed
to a 45° slope using a long-handled rake, and covered by at least 150 mm of soil excavated from
the next trench. The width of the trenches should not be less than their depth
Source: After Flintoff (1984)

estimated that 2% of the population of Medellín, Colombia, lives on income from the recycling of the city's refuse. Scavengers may be organized into co-operatives and provided with facilities at transfer points or at the tip, including water and electricity for cleaning and recycling purposes, and sewing machines to make items such as mops and mattresses. Scavengers may be trained in carpentry, metal-work, and other skills necessary to make better use of scrap materials.

Of special interest from a public health viewpoint is the control of seepage from tips. Over half the rain falling on a tip will probably be absorbed or evaporated. The remainder will flow out of the base and this 'leachate' may contain a high concentration of bacteria (up to 10^8 coliforms/100 ml), BOD (up to 30 000 mg/l), nitrogen, sulphides, and chlorides. The leachate may also contain any toxic or poisonous substances found in the tip. The quality of the leachate tends to improve as the tip becomes older. The danger to ground and surface water abstraction points depends completely on the local topography, geology, and climate. Advice, particularly from hydrogeologists, should be taken before a tip is located. The volume of leachate may be reduced by drainage of the tip surface, by diverting runoff from adjacent areas, and by planting vegetation on the top of the tip.

Incineration

Refuse may also be burned. An advantage of incineration is that it can be carried out relatively close to the centre of a city, thus saving in transport costs. However, efficient incineration requires sufficiently combustible refuse, and a fairly expensive incineration plant which is reliably operated. A common problem is the difficulty of keeping collected refuse dry in wet weather. Even in dry weather, the refuse collected in many tropical cities requires large (and prohibitively expensive) amounts of fuel oil for its incineration. It is still necessary to dispose of the residue, normally by tipping. Although its volume and weight is greatly reduced by incineration, most plant nutrients in the refuse have been lost so that the residue is of negligible value to agriculture.

Composting

The major alternative to tipping and incineration is composting. Flintoff (1984) reports that 88% of the contents of Indian refuse are acceptable for composting. The figures from Mexico and Britain are 65% and 64% respectively. Composting converts the organic content of refuse into compost, a soil conditioner which can improve the fertility and structure of agricultural soils. Its agricultural value is greatest if refuse is composted together with suitable quantities of nightsoil or sludge. The sale of compost to farmers can help to offset the running costs of a town's waste disposal system. Nearly two-thirds of Indian cities and thousands of villages in China dispose of some or all of their refuse by composting it with nightsoil. Composting is further discussed in Chapter 13.

Choice of method

Although other factors, particularly the public health aspects, cannot be ignored when choosing a refuse disposal system for a particular city, in practice the determining factor is usually its cost. Refuse disposal costs per ton in developing countries are not much less than in the industrialized world, and represent a heavy burden on meagre municipal budgets.

When offsetting the expected income from sale of compost against the higher cost of composting, over-optimistic forecasts should be avoided. Economies of scale in refuse treatment and handling plants are small, and can be approximated by the formula:

$$\text{(capital cost)} = \text{(constant)} \times \text{(capacity)}^{0.8}$$

Care should be taken, when comparing alternative systems, to include all the costs of transport, transfer etc., and to compare like with like.

12.6 REFERENCES AND FURTHER READING

Cointreau, S.J. (1982). *Environmental Management of Urban Solid Wastes in Developing Countries; a Project Guide*, Urban Development Technical Paper No. 5 (Washington DC: World Bank).

Flintoff, F. (1984). *Management of Solid Wastes in Developing Countries*, WHO Regional Publications, South-East Asian Series No. 1, 2nd edition (New Delhi: World Health Organization).

Gotaas, H.B. (1956). *Composting*, Monograph Series No. 31 (Geneva: World Health Organization).

Hagerty, D.J., Pavoni, J.L. and Heer, J.E. (1973). *Solid Waste Management* (New York: Van Nostrand Reinhold).

Holmes, J.R. (1984). *Managing Solid Wastes in Developing Countries* (London: John Wiley).

James, S.C. (1977). Metals in municipal landfill leachate and their health effects, *American Journal of Public Health*, **67**, 429–432.

Pickford, J. (1977). Solid waste in hot climates, *in Water, Wastes and Health in Hot Climates*, Feachem, R., McGarry, M., and Mara, D.D. (eds) (London: John Wiley), pp. 320–344.

Suess M.J. (1985). *Solid Waste Management: Selected Topics* (Copenhagen: WHO Regional Office for Europe).

Thanh, N.C., Lohani, B.N. and Tharun, G. (1978). *Waste Disposal and Resources Recovery* (Bangkok: Asian Institute of Technology).

UNCHS (Habitat) (1990). *Refuse Collection Vehicles for Developing Countries* (Nairobi: UN Centre for Human Settlements).

13

Composting

13.1 AEROBIC AND ANAEROBIC COMPOSTING

The composting of organic refuse with nightsoil or sewage sludge is a method of preparing them for application to the land as fertilizer and soil conditioner, rendering them more beneficial to the soil while killing any pathogens present. Composting may be aerobic or anaerobic. Different species of bacteria are responsible in each case, different chemical changes take place, and different temperatures are reached.

Anaerobic composting or digestion, also known as fermentation or putrefaction, takes place in the absence of air or oxygen. It is the process by which organic muds are broken down by bacteria in marshes, producing marsh gas, consisting mainly of methane (CH_4), but also of ammonia (NH_3) and various sulphur-containing gases, which smell unpleasantly. A small amount of warmth is produced, but more of the energy is stored in the methane gas, which may be used for cooking or heating.

Anaerobic composting is slow and unreliable as a method of pathogen destruction because it does not achieve sufficiently high temperatures. One method, the Bangalore method, requires alternate layers of refuse and nightsoil to be buried in trenches and left for at least six months.

In *aerobic* composting, oxygen-using micro-organisms feed on the organic matter and multiply. This is the decomposition process which occurs naturally on a forest floor when droppings from trees and animals are converted by micro-organisms into humus. Aerobic composting is normally odour-free. Aerobic composting typically produces twenty times as much heat as putrefaction. The centre of a well-aerated compost pile can reach temperatures of over 65°C. It is important to keep the moisture content of the compost between about 40 and 60%. If it is too wet it cannot contain enough air, but if it is too dry it will not conserve heat well enough to warm up, nor provide a suitable medium for bacterial growth. The moisture content can be adjusted by frequent turning if it is too high, and by adding water if it is too low.

The chief advantage of aerobic over anaerobic composting is its higher temperature, which causes greater speed of digestion and

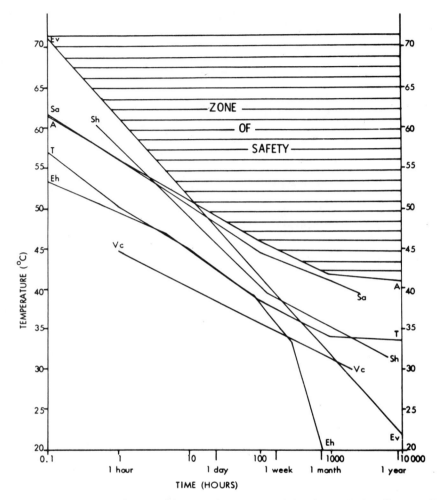

Figure 13.1 The influence of time and temperature on a variety of excreted pathogens. The lines drawn represent conservative upper boundaries for death

Ev	Enteroviruses	Eh	*Entamoeba histolytica cysts*
Sa	*Salmonella*	A	*Ascaris eggs*
Sh	*Shigella*	T	*Taenia eggs*
Vc	*Vibrio cholerae*		

Note: These lines are plotted from the graphs derived in Feachem *et al.* (1983). The lines represent very conservative estimates of the time–temperature combinations required to inactivate various pathogens. A process with time–temperature characteristics lying within the 'zone of safety' should guarantee the inactivation of all excreted pathogens, with the possible exception of hepatitis A virus at short times. Suitable time–temperature properties include:

at least 62°C for 1 hour;	at least 43°C for 1 month;
at least 50°C for 1 day;	at least 42°C for 1 year; .
at least 46°C for 1 week;	

effective pathogen destruction. Each group of excreted pathogens has a different ability to withstand raised temperatures for various lengths of time. These properties are reviewed in detail by Feachem *et al.* (1983) and some of the data are summarized in

Figure 13.1. The toughest excreted pathogens are the enteroviruses and *Ascaris* eggs. The lines plotted in Figure 13.1 are conservative and time–temperature combinations within the 'zone of safety' should guarantee destruction of nearly all excreted pathogens. The toughness of *Ascaris* eggs has caused them to be adopted in China as the indicator of compost quality, as *E. coli* alone are an inadequate indicator for this purpose.

13.2 TECHNIQUES

A variety of more or less mechanized techniques exists for composting. A widely applicable technique is to form long mounds, called windrows. These may comprise alternate layers of refuse and nightsoil (or sludge) and are typically about 1.5–2 m in height (Figure 13.2). The windrows are turned (either by hand or by tractor-shovel) every few days, to keep them aerobic and to ensure that all parts of the pile spend some time at the hot centre. The top of each windrow is shaped to throw off the rain.

An alternative is to place the compost in drained, cement-lined pits about 1 m deep, stacking it to 0.3 m above ground level. Pits are easier to load without spillage, and because they conserve heat better they do not need so much turning. The compost can be digested aerobically at first in order to raise its temperature and kill pathogens, and allowed to cool slowly under increasingly anaerobic conditions. However, some turning is necessary to ensure that all parts of the heap undergo the high temperatures in the interior.

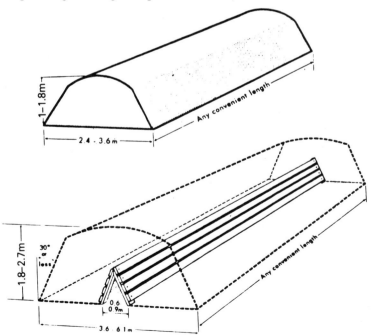

Figure 13.2 Two sizes and types of windrow *Source*: From Gotaas (1956)

Up to a quarter of the nutrient content of compost may leach from it in the liquor which seeps from the pile. This may be collected in drains and applied to the same or to other piles or pits, to keep them moist and conserve nutrients.

The frequency with which compost has to be turned will depend on local conditions. The best guide is to turn it when it has become anaerobic, as indicated by foul odours given off when it is disturbed with a shovel, or by a drop in temperature. The effect of turning in enabling the temperature to increase is illustrated in Figure 13.3, which also shows the effect of temperature on *Ascaris* eggs.

Turning a compost pile tends to reduce the degree of pathogen removal achieved by the process, because the parts of the pile which have been hot enough to kill micro-organisms can then be recontaminated by contact with those which have not. Turning also increases the cost and the amount of space required. Avoiding the need for turning can therefore be an advantage.

Another simple method which avoids the need for turning has been developed in recent years. It is particularly suitable for towns and cities. Known as the forced aeration method, it involves blowing air into the base of the pile through a 100 mm diameter perforated pipe (Figure 13.4). A 0.5 hp (37 W) air blower is more than adequate for a 50 tonne pile of material, which is typically 15 m × 4 m × 1.5 m high, with a triangular cross-section. The pipe is buried at the base of the pile in a layer of coarse material, such as

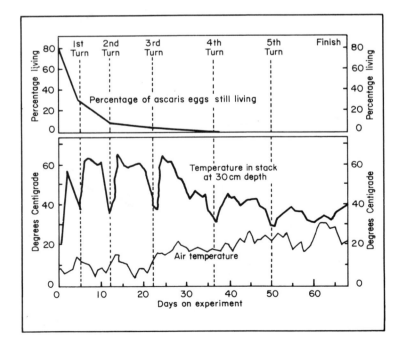

Figure 13.3 The relationship between windrow turning, temperature, and *Ascaris* egg survival in a compost stack *Source*:From Gotaas (1956)

Figure 13.4 Forced aeration composting. The method was originally developed by the US National Parks Service, sucking the air through a mixture of woodchips and sludge. *Blowing* the air reduces the odour and avoids the problem of condensate in the outlet pipe, and also helps to ensure high temperatures throughout the pile, and hence better pathogen removal
Source: Adapted from Shuval *et al.* (1980).

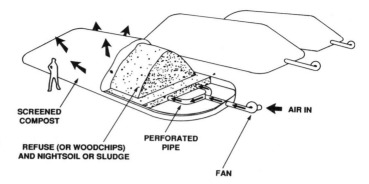

SCREENED COMPOST

REFUSE (OR WOODCHIPS) AND NIGHTSOIL OR SLUDGE

PERFORATED PIPE

AIR IN

FAN

straw. The pile is covered with 100–300 mm of screened compost, to retain heat and prevent fly breeding.

A thermocouple buried in the pile can be used to switch the blower on and off so as to keep the pile temperature in the range 50–60°C. Typically, it is on for about 20% of the time initially; this percentage increases to 75% during the first week, and then falls gradually to about 10% at the end of the pile's 30-day life. Finally, the compost is further improved by storage for two to four months.

Although the method was originally developed for mixtures of wood chips and dewatered sewage sludge, it has been adapted for nightsoil and organic refuse. Further details are given by Stentiford *et al.* (1985). More information on the composting of refuse and nightsoil may be found in Flintoff (1984) and Gotaas (1956).

13.3 CARBON/NITROGEN RATIO

The micro-organisms feeding on a compost heap use nitrogen, some of the carbon and other available nutrients to build cell protoplasm, but roughly twice as much carbon is used as a source of energy, producing carbon dioxide (CO_2) which is released. Much more carbon than nitrogen is required, or else some of the nitrogen, an important plant nutrient, may be lost to the air in ammonia.

On the other hand, if insufficient nitrogen is available, the organisms will begin to die when it has been used up, and their own nitrogen and carbon then become available to other organisms. Several generations of bacteria are therefore required to burn up the carbon. This slows down the reaction and prevents it from reaching such high temperatures. Moreover, if the final product is deficient in nitrogen, bacteria in the compost will extract nitrogen from the soil to complete the reaction, thus robbing the soil of fertility.

There is therefore an ideal ratio between the carbon and nitrogen content (the C/N ratio) of the raw compost, of about 30:1 to 40:1. The ratio can be adjusted by blending different ingredients such as straw, domestic refuse, animal manure, and nightsoil. Straw and other vegetable waste has a high C/N ratio, but nightsoil, and especially urine, brings the ratio down. Table 13.1 gives the nitrogen content and C/N ratios of various compostable materials.

Because of its low C/N ratio and its high moisture content, only a limited amount of nightsoil can be added to compost, typically about 20–30%. In many tropical cities, the weight of compostable refuse collected per person per day is about the same as the weight of nightsoil, so not all the nightsoil can be composted with it. However, it is sometimes possible to supplement domestic refuse with agricultural wastes. In rural areas, agricultural wastes are more easily available, although it may be necessary to substitute alternative sources of fuel where vegetable wastes or manure are used for cooking fires.

13.4 PROBLEMS OF COMPOSTING

Composting can be carried out with simple equipment and very little training, and is well suited to many tropical environments. It is particularly appropriate for small cities and towns near to agricultural areas. However, unforeseen problems have sometimes arisen when it has been introduced into new areas without prior experimentation.

First, it may be necessary to separate out glass, metal, and other non-biodegradable matter from the refuse. This may be done mechanically or by hand, before or after the composting process. Polythene bags have been a particular problem in some places; if spread on the land they are dangerous to livestock which may eat them. If it is not possible to sort all the refuse by hand, a cutter/shredder should be used to reduce the material to strips 30 mm wide and 150 mm long.

Table 13.1 Approximate nitrogen content and C/N ratios of some compostable materials (dry basis)

Material	N(%)	C/N
Urine	15–18	0.8
Poultry manure	6	—
Mixed slaughterhouse wastes	7–10	2
Nightsoil	6	6–10
Sheep and pig manure	4	—
Domestic vegetable refuse	2	25
Cow manure	2	—
Wheat straw	0.3	128
Sawdust	0.1	511

Source: From Gotaas (1956).

Second, it may not be easy to persuade farmers to use the new fertilizer, particularly if it includes composted nightsoil or has to compete with subsidized chemical fertilizers, or contains non-biodegradable materials such as splinters of glass, metals, or plastics. A price and a market for the compost should be assured before it is produced on a large scale.

A third problem is the breeding of flies. Organic refuse and nightsoil usually contain some fly eggs and larvae before they come to the composting site, but is it possible to prevent most of them from hatching by turning the compost early and often, for instance on the third and eighth days, taking care to place the surface material in the hot centre of the turned pile. Composting should be carried out on a hard surface, to prevent the larvae from escaping from the heat by burrowing into the soil beneath.

In order to ensure that the technique is workable and appropriate, to find the best combination of materials and turning times, and to be able to test-market the product, it is advisable to experiment with composting on a pilot scale before introducing it for a whole community.

13.5 INDIVIDUAL COMPOSTING TOILETS

The scientific principles which apply to the composting of municipal wastes also apply to individual composting toilets. In practice, these have hardly ever achieved aerobic conditions in a tropical setting, and anaerobic processes take many months (usually at least six months) to produce a compost which is relatively free of pathogens. Ideally, carbon/nitrogen ratios should be adjusted to initial levels of 30 to 40 by the addition of vegetable wastes, ash, or sawdust, and urine should be excluded or drained off; in practice, it is difficult to train the users to do this reliably so that composting toilets are not recommended for introduction to poor communities.

13.6 REFERENCES AND FURTHER READING

FAO (1977). *China: Recycling of Organic Wastes in Agriculture*, Soils Bulletin No. 40 (Rome: Food and Agriculture Organization).

Feachem, R.G., Bradley, D.J. Garelick, H. and Mara, D.D. (1983). *Sanitation and Disease: Health Aspects of Excreta and Wastewater Management* (London: John Wiley).

Flintoff, F. (1984). *Management of Solid Wastes in Developing Countries*, WHO Regional Publications, South-East Asian Series No. 1, 2nd edition (New Delhi: World Health Organization).

Gotaas, H.B. (1956). *Composting*, Monograph Series No. 31 (Geneva: World Health Organization).

McGarry, M.G. and Stainforth, J. (1978). *Compost, Fertilizer, and Biogas Production from Human and Farm Wastes in the People's Republic of China* (Ottawa: International Development Research Centre).

Shuval, H.I., Gunnerson, C.G. and Julius, D.S. (1981). *Nightsoil Composting*, Appropriate Technology for Water Supply and Sanitation No. 10 (Washington, DC: World Bank).

Stentiford, E.I., Taylor, P.I., Leton, T.G. and Mara, D.D. (1985). Forced aeration composting of domestic refuse and sewage sludge. *Journal of the Institute of Water Pollution Control*, **84**, 23–32.

Winblad, U. and Kilama, W. (1985). *Sanitation Without Water* (London: Macmillan).

14

Health Aspects of Waste Re-use

14.1 INTRODUCTION

Human wastes are a valuable natural resource and should not be thrown away (Figure 14.1). Re-use of sewage, nightsoil, organic refuse and sludge derived from sewage treatment processes is possible in most situations. As discussed in Chapter 8, there are three main forms of re-use: agriculture, aquaculture and biogas. There are no specific health risks associated with biogas production other than those due to the handling of excreta which apply to many sanitation systems. We will therefore discuss only those health risks related to agriculture and aquaculture.

Human wastes from all communities contain pathogens (Table 1.3). These pathogens survive to different degrees as the waste is transported, treated and applied to the land or pond. The health risk associated with waste re-use depends on the degree of treatment which has been provided and the nature of the re-use process. A detailed account of this subject is provided by Mara and Cairncross (1989).

14.2 HEALTH AND AGRICULTURAL RE-USE

Sewage and nightsoil are often used in agriculture, sometimes in an organized way but often informally, illegally or clandestinely. In arid areas, such as coastal Peru and the middle East, sewage is at a premium because water is scarce, and in densely populated parts of the world, such as China and India, excreta are valued as a fertilizer.

The river water used for irrigation in many developing countries, and also in Europe and North America, often contains a substantial percentage of municipal sewage. This can also be considered an indirect form of re-use.

Many re-use schemes are found in industrialized countries. For example, the Werribee farm irrigated with treated sewage from the

The Good Daughter-in-law
by Hsieh Chang-yi

"Early in the morning, the magpies cry,
The newly-wed daughter-in-law is carrying excreta on a pole
Liquid from the excreta stains her new trousers
The hot sweat soaks into her embroidered jacket
The commune members praise her and mother is pleased
All tell her she has got a good daughter-in-law."

Figure 14.1 In some countries, such as China, the use of human excreta in agriculture is an ancient practice and may provide the main source of fertilizer for the fields. In other countries, such as much of Africa, it is considered much less acceptable to handle human excreta. In China, cartage systems (such as that in Figure 8.9) remove excreta from houses and transport them by hand, truck or boat to agricultural areas where they are applied to the fields. Hygienic problems arise if the excreta contaminate the workers (as in the poem) or if excreta are applied raw to the fields (as is common in China, now that most of the commune composting brigades have been disbanded) *Source*: From McGarry and Stainforth (1978). Reproduced by permission of the International Development Research Centre

city of Melbourne, Australia, supports a herd of 13 000 cattle and up to 30 000 sheep. Wastewater and sludge from the city of Braunschweig in Germany have been used since 1971 to irrigate 2 800 ha of land belonging to a number of farmers. Since 1984 in the town of Kearney, Nebraska in the USA the sludge from the wastewater treatment plant is mixed with animal manure and composted, before spreading on 1200 ha of land on which maize is grown; the increased humus content improves the water retention of the sandy soil, and the nutrients in the compost avoid the need for artificial fertilizer, except for supplemental nitrogen.

These and other cases show that wastes *can* be used safely. However, it is also clear that the uncontrolled use of untreated human wastes can pose a health risk, especially to farm workers and crop handlers, but also to the consumers of some edible crops.

In the past, it was held that if any pathogen reached the fields it constituted a potential risk to health. It followed that the only safe precaution was to require treatment of the wastes to remove *all* pathogens, so that this could not occur. This led to the setting of treatment standards which were practically impossible to achieve. When these were observed, use of the waste was not worthwhile economically. More often, they were ignored and the waste was used informally, with no treatment at all.

A 'potential risk' may not necessarily become an 'actual risk'. Pathogens reaching a field may not contaminate crops (for instance, if fruit trees are irrigated at ground level) or may not survive long enough or in sufficient numbers to infect people; and the population may be immune or take other measures to avoid infection, such as wearing protective clothing. Moreover, a particular infection may have other routes of transmission in the community, so that some of the disease observed may not be associated with waste re-use. The most useful measure of risk, therefore, is the '*attributable risk*' or 'excess risk', which is a measure of the amount of disease associated with waste re-use.

Consideration of attributable risk represents a change from microbiological to epidemiological criteria. Its assessment requires epidemiological studies comparing the health of two populations, one exposed to the wastes and one which is not. Reviews of existing epidemiological studies (Feachem and Blum, 1985; Shuval *et al.*, 1986) have led to a reassessment of the health risks of waste re-use in agriculture, with two main conclusions.

First, full treatment of the wastes is not the only way to protect the health of workers and consumers. Other measures can be used to protect these groups from infection (Figure 14.2). Briefly they are the following:

- *Application* Careful choice of the method of application of the wastes can help to control the risk. For example, localized (trickle, drip or bubbler) irrigation reduces both the risk of crop contamination and the exposure of workers to the wastewater; application of excreta before planting is less risky than application during the growing cycle; it will also help to reduce risks if wastewater irrigation ceases several weeks before the harvest.
- *Crop restriction* For example, consumers will be safe from infection if industrial crops such as cotton, sisal, or timber are grown; animal fodder crops, or grains for processing into flour, will also reduce the risk to human consumers.
- *Human exposure control* This includes a number of measures such as wearing protective clothing, immunization and de-worming of farm workers, and also milk pasteurization and meat inspection where pasture or fodder crops are grown using the waste.

Judicious combinations of these, often with partial treatment, can be sufficient to protect the health of all concerned. Some such combinations are shown in Figure 14.3.

Second, the highest risk from the use of untreated wastes relates to infection with intestinal nematode worms, particularly *Ascaris*, *Trichuris* and hookworm. This affects farm workers, whatever crop is grown, and also consumers of some edible crops. There is generally a lower attributable risk from bacterial and protozoal enteric infec-

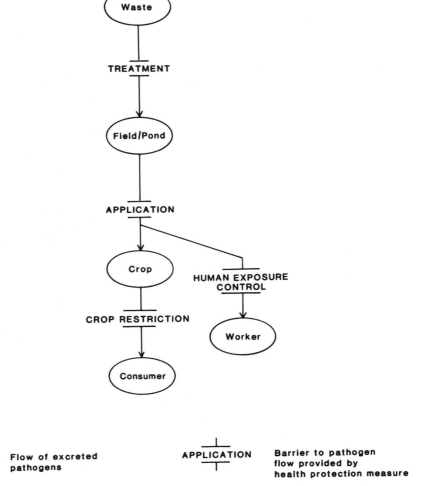

Figure 14.2   Effect of health protection measures in interrupting the potential transmission routes of excreted pathogens

tions, and less still from viruses. Because of this, the WHO quality guideline for treated wastewater to be used for irrigation was made stricter by requiring a maximum of 1 nematode egg per litre. On the other hand, the bacterial standard was relaxed to allow a geometric mean of 1000 faecal coliforms per 100 ml (Table 14.1). Both of these guideline values can be achieved by a well-designed system of stabilization ponds (see Chapter 10), or by upgrading a conventional treatment plant through the addition of one or more ponds.

For sludge and nightsoil, composting is generally the appropriate treatment process, and should preferably be aerobic (see Chapter 13). Twelve months may be necessary to remove the hardy pathogens

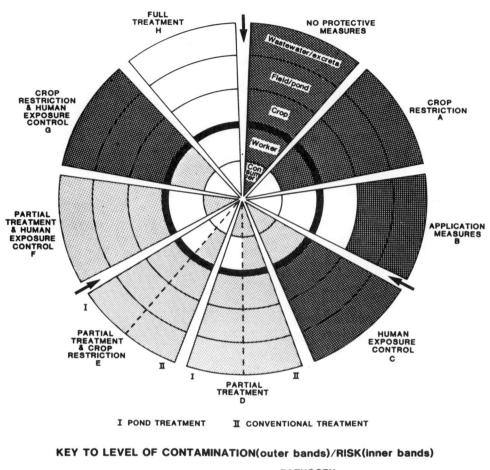

FULL TREATMENT H

NO PROTECTIVE MEASURES

Wastewater/excreta

Field/pond

Crop

Worker

Consumer

CROP RESTRICTION & HUMAN EXPOSURE CONTROL G

CROP RESTRICTION A

PARTIAL TREATMENT & HUMAN EXPOSURE CONTROL F

APPLICATION MEASURES B

I

PARTIAL TREATMENT & CROP RESTRICTION E

II

I

II

HUMAN EXPOSURE CONTROL C

PARTIAL TREATMENT D

I POND TREATMENT   II CONVENTIONAL TREATMENT

KEY TO LEVEL OF CONTAMINATION(outer bands)/RISK(inner bands)

HIGH   LOW   SAFE   **PATHOGEN FLOW**   **BARRIER**

Figure 14.3   Generalized model to show the level of risk to human health associated with different combinations of control measures for the use of wastewater or excreta in agriculture or aquaculture

(notably *Ascaris* eggs) from an anaerobic compost pile, but forced aeration composting can kill them in one month.

There is some concern in the industrialized countries over the presence of heavy metals in urban sewage sludge used as fertilizer, particularly lead, mercury, zinc, and most importantly cadmium. These metals are less likely to be present in the developing countries, but they are not entirely removed by sewage treatment works, tend to accumulate in the soil over the years, and can be taken up by cereals and other crops. In many cases, the concentration of toxic chemical pollutants in wastewater is reduced by about 50% by sedimentation.

Table 14.1   Recommended microbiological quality guidelines for wastewater use in agriculture[1]

Category	Reuse conditions	Exposed group	Intestinal nematodes[2] (arithmetic mean[3] no. of eggs per litre)	Faecal coliforms (geometric mean[3] no. per 100 ml)	Wastewater treatment expected to achieve the required microbiological quality
A	Irrigation of crops likely to be eaten uncooked, sports fields, public parks[4]	Workers, consumers, public	≤ 1	$\leq 1000^4$	A series of stabilization ponds designed to achieve the microbiological quality indicated, or equivalent treatment
B	Irrigation of cereal crops, industrial crops, pasture and trees[5]	Workers	≤ 1	No standard recommended	Retention in stabilization ponds for 8–10 days or equivalent helminth and faecal coliform removal
C	Localized irrigation of crops in category B if exposure of workers and the public does not occur	None	Not applicable	Not applicable	Pretreatment as required by the irrigation technology, but not less than primary sedimentation

[1]In each specific case, local epidemiological, sociocultural and environmental factors should be taken into account, and the guidelines modified accordingly.
[2]*Ascaris* and *Trichuris* species and hookworms.
[3]During the irrigation period.
[4]A more stringent guideline (≤ 200 faecal coliforms per 100 ml) is appropriate for public lawns, such as hotel lawns, with which the public may come into direct contact.
[5]In the case of fruit trees, irrigation should cease two weeks before fruit is picked, and no fruit should be picked off the ground. Sprinkler irrigation should not be used.

Since the volume of settled sludge is about one-half of one per cent of the volume of sewage, each constituent that is transferred to the sludge is then concentrated about a hundredfold. Where heavy metals are present in sludge in a significantly higher concentration than in animal manure, the quantity of sludge to be applied should be limited. In Sweden, for example, it is recommended that not more than 1 t/ha.year (dry weight) of sludge should be applied to the land.

14.3 HEALTH AND FISH FARMING

Aerobic ponds containing sewage or nightsoil typically support large growths of algae. These may be used to support populations of fish, which can be harvested for consumption. For example, 44 km^2 of ponds, fed with raw wastewater from Calcutta, produce over 1 ton/ha of fish (carp and tilapia) each year, amounting to 10–20% of all the fish consumed by the city's 11 million inhabitants. Three health problems are potentially associated with the use of excreta to fertilize fish ponds.

(1) *Passive transference of pathogens by fish* Although fish do not suffer from any human bacterial pathogens, if faecal bacteria are very numerous in the pond water they may accumulate on the surfaces or in the intestines of the fish. WHO (1989) has therefore suggested a tentative bacterial quality guideline for the pond water of not more than 1000 faecal coliforms/100 ml. The level in the wastewater entering the pond can be several times greater than this, since the wastewater is usually diluted in the pond. Keeping the pond water quality within the guideline will also help to protect the health of the fishery workers who are in contact with it.

Keeping the fish in clean water for 2 to 3 weeks before harvest, known as 'depuration', will further reduce any faecal contamination of the fish and also remove any residual objectionable odours. A promising alternative is to grow 'trash' fish such as tilapia in waste-fertilized fish ponds and then feed them to high-value fish.

(2) *Transmission of some helminths for which edible fish, shellfish, crustaceans or plants are intermediate hosts in the life cycle* The most important of these are the Chinese liver fluke *Clonorchis sinensis*, cat liver flukes *Opisthorchis* spp., the fish tapeworm *Diphyllobothrium latum*, and the intestinal fluke *Fasciolopsis buski*. The first three of these infect man when insufficiently cooked fish are eaten, and the last is found on edible water plants such as water chestnut. All of them are only found in certain parts of the world, particularly South-east Asia and the Far East (Figure 14.4). Their life cycles are listed in Appendix C, and distribution maps are given by Feachem *et al.* (1983).

In countries where these parasites are not found, they are unlikely to be introduced by aquaculture alone, as all of these flukes depend for their transmission on specific species of snail. Where they are endemic, the wastes should be treated to kill any of their eggs which might be present. For example, storage of nightsoil or sludge for 7 days before addition to the ponds is perfectly adequate to remove *Clonorchis* and *Opisthorchis* eggs. However, the effect of waste-fed fish ponds on the transmission of these parasites is often minor in comparison with other bodies of water where fish or water plants are taken for consumption, which may be contaminated by indiscriminate defecation, by the use of overhang latrines, or by animals,

Figure 14.4 Raw nightsoil is poured into fishponds in Taiwan. This practice is widespread in East and South-east Asia, and greatly contributes to fish production and thus to dietary protein. However, major health hazards are involved unless the nightsoil is first pre-treated to remove various pathogens (Photo: M McGarry)

since the eggs of all of the parasites concerned can also be found in the faeces of vertebrate animals other than man.

(3) *The risk to fishery workers of schistosomiasis* This is only a concern in those parts of the world where one form or another of schistosomiasis is endemic (see Chapter 17), and where the local vector species of snail is likely to breed in fish ponds. Certainly, fishermen are at increased risk of schistosomiasis in many communities in Africa and China.

Storage of the wastes before adding them to ponds is an effective treatment measure. In China, storage of faeces with urine in warm weather killed schistosome eggs in 3 days by ammonia toxicity; adding urea, ammonium bicarbonate and calcium cyanamide, at 1%, 0.5% and 0.25% respectively, kills them in 24 hours (Edwards 1992). These measures are likely to be more effective in Africa or Brazil, because the eggs of *Schistosoma mansoni*, which is endemic there, are less robust in the environment than those of *S. japonicum*, the species found in the Far East.

Removal of aquatic vegetation from the banks of a pond makes it more likely (although not certain) that it will not harbour snails. However, biological control with snail-eating fish has not proved promising; large numbers of snails have been found in dense patches of vegetation, presumably where the fish could not reach them. The tilapia *Oreochromis mossambicus*, which is a predator of snail eggs and young, seems to offer more hope.

Other forms of exposure control, such as chemotherapy and protective clothing for the pond workers, are also appropriate, and feasible when the operation is carried out on a limited scale such as a set of waste stabilization ponds. The success of environmental schistosomiasis control efforts in China, a country where excreta-fed fish ponds are common, suggests that the problem can be kept to manageable proportions by good pond management.

14.4 REFERENCES AND FURTHER READING

Edwards, P. (1992). *The Reuse of Human Wastes in Aquaculture: a Technical Review*, Water and Sanitation Report No. 2 (Washington DC: The World Bank).

Edwards, P. and Pullin, R.S.V. (1990). *Wastewater-fed Aquaculture; Proceedings of an International Seminar* (Bangkok: Asian Institute of Technology).

Feachem, R.G. and Blum, D. (1985). Health Aspects of Wastewater Reuse, in *Reuse of Sewage Effluent* (London: Thomas Telford).

Feachem, R.G., Bradley, D.J., Garelick, H. and Mara, D.D. (1983). *Sanitation and Disease: Health Aspects of Excreta and Wastewater Management* (London: John Wiley).

McGarry, M.G. and Stainforth, J. (1978). *Compost, Fertilizer and Biogas Production from Human and Farm Wastes in the People's Republic of China* (Ottowa: International Development Research Centre).

Mara, D.D. and Cairncross, S. (1989). *Guidelines for the Safe Use of Wastewater and Excreta in Agriculture and Aquaculture* (Geneva: World Health Organization).

Shuval, H.I., Adin, A., Fattal, B., Rawitz, E. and Yekutiel, P. (1986). *Wastewater Irrigation in Developing Countries: Health Effects and Technical Solutions*, World Bank Technical Paper No. 51 (Washington DC: The World Bank).

Strauss, M. and Blumenthal, U.J. (1990). *Use of Human Waste in Agriculture and Aquaculture: Utilization Practices and Health Perspectives*, IRCWD Report No. 08/90 (Duebendorf, Switzerland: International Reference Centre for Wastes Disposal).

WHO (1989). *Health Guidelines for the Use of Wastewater in Agriculture and Aquaculture*, WHO Technical Report Series No. 778 (Geneva: World Health Organization).

Part IV

Environmental Modifications and Vector-borne Diseases

15

Engineering Control of Arthropod Vectors

15.1 INTRODUCTION

Certain diseases are transmitted by particular species of arthropod, known as vectors. The most important arthropod vectors are mosquitoes, flies, bugs, ticks and lice. Vectors may be mechanical or biological. A *mechanical* vector simply transports pathogens on or in its body from one place to another. An example is the transportation of faecal pathogens by flies or cockroaches. This chapter deals mainly with biological vectors. A *biological* vector is actually infected by the pathogen, which develops or multiplies (or both) inside the body of the vector. Any disease having such a vector is clearly open to control by engineering measures directed against that vector. Table 15.1 gives some generalized information about the vectors of the major vector-borne diseases, which include many serious tropical infections.

15.2 MOSQUITO-BORNE DISEASES

Mosquitoes are vectors of malaria, filariasis, and arboviral infections. Only female mosquitoes bite; they need the blood meal to feed their developing eggs until they are ready to lay. Gravid, blood-laden mosquitoes cannot fly very far, so they generally bite within a kilometre or so of their breeding place. Female mosquitoes may take blood from all warm-blooded, and some cold-blooded, land animals, but each mosquito species has marked preferences for particular sources of blood. Many species bite man and some of these are vectors of human diseases. Each species, or subspecies, has a particular ecology, and control measures must be specific to the target mosquito.

Mosquitoes lay their eggs in water, but different species have differing, and often very specific, requirements for a suitable site. Figure 15.1 shows how the two major families of mosquito, the

Table 15.1   Generalized biological information concerning the vectors of some diseases

Disease	Vector of intermediate host	Reproductive potential				Feeding time	Preferred behaviour		
		Number of eggs	Egg-to-egg cycle	Number of broods	Life-span		Resting place	Source of blood	Flight or dispersal range[1]
Malaria	*Anopheles* mosquitos	200	10–14 days	6–10	20 weeks	Night	Indoors and outdoors	Man and animals	1.5 km
Filariasis; viral diseases	*Culex* and *Aedes* mosquitos	200	8–10 days	6–10	20 weeks	Night and day	Indoors and outdoors	Man and animals	0.1–8 km
Onchocerciasis	Blackflies	400	2–3 weeks	3–4	1–2 weeks	Day	Outdoors	Man and animals	4–8 km
African trypanosomiasis	Tsetse flies	1 pupa	60 days	10	3–12 weeks	Day	Outdoors	Man and animals	2–4 km
Leishmaniasis	Sandflies	50	6–8 weeks	2	12 weeks	Night	Outdoors	Man and animals	50 m
Chagas' disease	Triatomid bugs	200	52 weeks	1–2	50 weeks	Night	Indoors	Man	10–20 m
plague {	Rats	8 per litter	12 weeks	4	32 weeks	Night and day	Indoors and outdoors	—	50–80 m
plague {	Fleas	12	8 weeks	10	15 weeks	Night and day	Indoors and outdoors	Animals and man	—

[1] Under normal, static conditions.

Note: The figures are only indicative and illustrate the major factors affecting transmission. They vary widely from species to species and in different environments.

Source: After WHO (1980).

Anophelines   Culicines

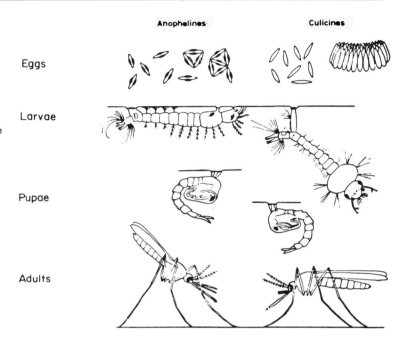

Eggs

Larvae

Pupae

Adults

Figure 15.1 The anopheline and culicine mosquitoes can be distinguished readily by the shape and clustering of their eggs, by the angle at which the larvae hang from the water surface, and by the angle of repose of the adults on a wall or other surface *Source*: From Busvine (1980). Reproduced by permission of British Museum (Natural History)

anophelines and the culicines, may be distinguished at each stage of their development.

Following the discoveries late in the nineteenth century of the role of mosquitoes in disease transmission (for instance by Manson working on filariasis in China and by Ross working on malaria in India) attention focused on destroying the breeding sites either by drainage or by covering suitable pools with oil. Ross showed that malaria could be eradicated from an area by greatly reducing the mosquito population, which is easier than eradicating the vectors completely.

In the 1930s the concept of *species sanitation* was developed. There might be many species of mosquito in a given environment, each with different habits; but species sanitation involved the identification of the principal vector species and its breeding habits, and the implementation of control measures aimed at eliminating breeding places with the very specific characteristics which that species required. Thus shade-loving breeders were controlled by clearing vegetation and sun-loving species were discouraged by planting. Salt-water breeders were controlled by engineering works to prevent salt-water intrusion into coastal lagoons and swamps. These engineering methods were highly successful in some areas, particularly in Asia, and formed the backbone of antimalarial and antifilarial programmes.

In the 1950s and 1960s these methods were almost completely superseded by the use of DDT and other residual insecticides. These were sprayed on the walls of houses and killed the resting adult mosquitoes. More recently, however, the increased price of insecticides, together with the increasing commonness of insect strains which are resistant to insecticides, has led to a re-examination of environmental methods for vector control, and these methods are likely to become increasingly important in the future.

Malaria

Malaria is a disease transmitted from person to person by certain species of mosquito of the genus *Anopheles* (Figure 15.1 and Table 15.2), and causes acute bouts of fever which recur at intervals. Although malaria has been eradicated in some countries it is still a major public health problem in many parts of the world (Figure 15.2). Malaria in man is caused by four species of protozoal parasite:

> *Plasmodium falciparum* — causing falciparum malaria especially in the humid tropics where transmission is possible all year round; this is the most serious form, and may be fatal.
> *Plasmodium vivax* — causing vivax malaria especially where, due to a pronounced dry or cool season, transmission is seasonal.
> *Plasmodium malariae* — causing quartan malaria, in which there are usually bouts of fever every three days; it has a patchy distribution in the tropics and subtropics.
> *Plasmodium ovale* — causing ovale malaria; uncommon and found mainly in West Africa.

By far the most common and serious malarial infections are caused by *P. falciparum* and *P. vivax*.

Three main types of control measure are in use: (i) administration of drugs to infected people; (ii) killing of adult mosquitoes; (iii) control of mosquito breeding by eliminating or treating suitable habitats. These are considered in turn.

(1) *Chemotherapy of infected people* Since man is effectively the only sufferer from human malaria, it is conceivable that if all malaria sufferers in an area could be cured, then no more mosquitoes could become infected and transmission of the disease would cease. However, it is usually impossible in practice to reach, in sufficient time, all the inhabitants of an area, or even of a small village. Moreover, while treatment may remove the symptoms of the disease, it does not necessarily eradicate the parasites from the bloodstreams or livers of all carriers. In addition, the situation is often complicated by migration of infected people in and out of a control area. The routine administration of drugs, for instance in salt, has been suggested, but there is a danger in the long run of this

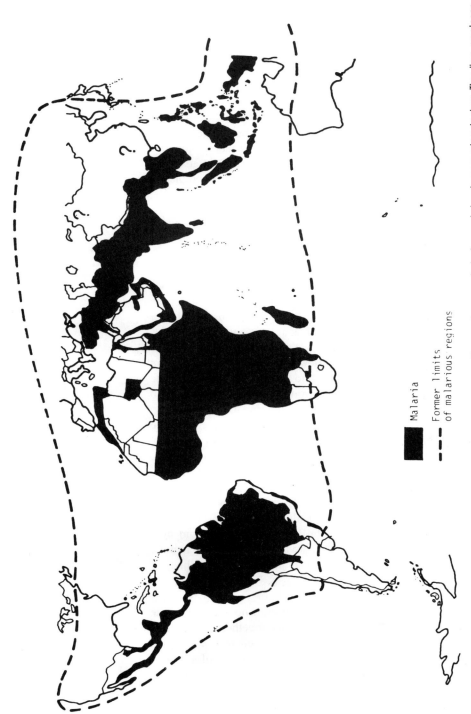

Figure 15.2   Distribution of malaria, past and present. Black areas are those where malaria transmission is occurring today. The line encloses areas where malaria transmission occurred historically. While malaria has been eliminated from several temperate areas, notably North America and Europe, little progress has been made in the tropics and malaria control in Africa has been particularly unsuccessful
Source: Data from WHO and Busvine (1975)

exacerbating the increasing problem of *Plasmodium* resistance to these antimalarial drugs.

Research is currently under way to develop and test a malaria vaccine, but it is likely to be many years before a cheap and effective vaccine is widely available.

(2) *Killing of adult mosquitoes*   This is usually accomplished with insecticide applied to the walls of houses. It has successfully reduced the incidence of malaria in some areas; but even where it has succeeded in eradicating malaria the mosquitoes remain and the malaria could be re-introduced. The insecticide requires repeated application. In some countries, such as Sri Lanka, the favourable picture of the mid-1960s has now been completely overtaken by a resurgence of malaria and malarial vectors. The use of DDT, and other residual insecticides, is becoming increasingly disfavoured due to environmental concern (largely in the USA and Europe) and to the soaring price of these products. The widespread and increasing resistance of *Anopheles* mosquitoes to insecticides is also a major problem.

A promising approach developed in recent years is to use bed nets treated with insecticide, usually a pyrethroid. This cheap measure renders the nets effective even if they have holes or tears. However, it is only effective against mosquito species which bite indoors and at night.

(3) *Treatment or elimination of breeding sites*   An egg laid in water by a mosquito develops into a larva and then a pupa (Figure 15.1) before emerging from the water as an adult mosquito five to fourteen days later. During this time it can be killed by chemical larvicides, which have frequently been used for this purpose, care being taken to avoid polluting water supplies. The use of larvicides is subject to the same qualifications as the use of insecticides to attack adult mosquitoes. Mineral and vegetable oils may be used, though they must be applied more frequently and in greater quantities than most insecticides. Oils (and also polystyrene beads; see below) work by discouraging the female mosquito from egg-laying and by suffocating the larvae. Rather than treating breeding sites, they may be eliminated by draining, filling, or changing crucial properties.

None of these approaches to malaria control is normally adequate on its own. They should preferably be applied together, in a mixture of methods chosen to suit local conditions. The control of breeding sites is the approach of most interest to the engineer, and a variety of methods is available for application to different types of site.

Drainage and filling, the most common methods, aim to remove standing water. Filling is particularly useful for the many small (frequently man-made) sites in or near towns and villages, such as borrow pits and water holes, although it can also be used on a larger scale. In coastal areas, for instance, dredging spoil can be used as fill material to reclaim swampy land.

Drainage can also enable marshy land to be reclaimed for productive use. An antimalarial drainage system need not necessarily be as ambitious or as expensive as agricultural or stormwater drainage. It is not usually necessary to remove all storm water immediately; only to ensure that no surface water persists for more than a few days after rainfall, so that any eggs laid in it have no time to develop.

A body of standing water too large to drain or fill can be treated by filling and deepening, so as to steepen the sides with material dredged from the bottom and thus to eliminate extensive areas of shallow water. This shortens the shoreline and limits the growth of emergent vegetation.

An important vector of malaria in urban India is *Anopheles stephensi*, breeding in the water storage tanks on buildings' roofs. Vector breeding in water cisterns has also been documented in East Africa. In such closed breeding sites, the polystyrene bead method, described below for filariasis control, may be appropriate.

Other methods include: coastal measures to change the salinity of breeding sites; the paving of the shoreline on ponds with constant water level (Figure 10.5); damming to impound water over a breeding area (though this needs care—see Chapter 16); and the periodic flushing of small streams (section 16.6).

Control measures should be concentrated within 1–2 km of human settlements. The siting of settlements away from mosquito-infested areas is the oldest of all control methods. Setting up a 'dry belt' or buffer zone of dry crops around each village in a wet rice-cultivation area can also be effective. If it is used for pasture, the presence of livestock may divert the vector from human to animal blood meals.

Within a town or village, an efficient drainage system helps to control malaria as well as other vector-borne diseases if it is well maintained, and the planting of water-avid trees such as eucalyptus may help to lower the water table. Vector control work is particularly important in the dry season, to prevent year-round transmission. Fortunately, this is the time when municipal road and drain maintenance gangs have relatively few other tasks.

Mosquito-control programmes which depend on mobilization of the population are not likely to succeed unless they are effective against several species, and thus have an immediately perceptible effect on the mosquito nuisance.

A good knowledge of the vector species and its behaviour is important, to avoid unnecessary work and to ensure that environmental modifications do not favour some other potential vector (Table 15.2). In parts of Africa and India, for instance, puddles formed in the bed of a drainage channel could be a better

Table 15.2   Twelve epidemiological zones of malaria and some of the important vectors

Zone		Extension	Main malaria vectors
1	North American	From the Great Lakes to southern Mexico	*A. quadrimaculatus, A. freeborni* with *A. albimanus* as an incidental vector
2	Central American	Southern Mexico, the Caribbean islands, fringe of the South American coast	*A. albimanus, A. aquasalis, A. punctimacula A. darlingi, A. aztecus*
3	South American	Most of the South American continent irregularly beyond the Tropic of Capricorn	*A. darlingi, A. aquasalis, A. pseudopunctipennis, A. bellator, A. cruzi*
4	North Eurasian	With the Palaearctic region excluding the Mediterranean coast of Europe.	*A. atroparvus, A. sacharovi, A. maculipennis, A. messeae.* To the east *A. pattoni, A.sinensis*
5	Mediterranean	Southern coast of Europe, north-western part of Africa, Asia Minor and east beyond the Arab sea	*A. labranchiae, A. sacharovi, A. superpictus, A. claviger, A. hispaniola, A. messeae*
6	Afro-Arabian	Africa north and south of the Tropic of Cancer including central part of the Arabian peninsula	*A. pharoensis, A. sergenti, A. multicolor, A. hispaniola, A. gambiae* (in part)
7	Afro-Tropical (formerly 'Ethiopian')	Southern Arabia, most of the African continent, Madagascar and the islands south and north of it	*A. gambiae* complex, *A. funestus, A. rufipes, A. moucheti, A. nili, A. pharoensis, A. d'thali*
8	Indo-Iranian	North-west of the Persian gulf and east of it including the Indian subcontinent	*A. sacharovi, A. superpictus, A. culicifacies, A. stephensi, A. fluviatilis, A. annularis*
9	Indo-Chinese Hills	A triangular area including the Indo-Chinese peninsula, the north-western fringe beyond the Tropic of Cancer	*A. minimus, A. leucosphyrus, A. balabacensis*
10	Malaysian	Most of Indonesia, Malaysian peninsula, Philippines and Timor	*A. leucosphyrus, A. balabacensis, A. sundaicus, A. maculatus, A. sinensis, A. umbrosus, A. aconitus, A. philippinensis. A. minimus flavirostris, A. barbirostris*
11	Chinese	Largely the coast of mainland China, Korea, Taiwan, Japan	*A. sinensis, A. pattoni, A. sacharovi*
12	Australasian	Northern Australia, the island of New Guinea and the islands east of it to about 175° east of Greenwich, but excepting the malaria-free zone of the south-central Pacific[a]	*A. koliensis, A. punctulatus, A. farauti, A. annulipes, A. bancrofti*

[a]The malaria-free zone of the south-central Pacific includes New Caledonia, New Zealand, the Caroline Islands, Marianas, up to Hawaian islands, east to Galapagos and Juan Fernandez and rejoining the southern tip of New Zealand.
Source: From Bruce-Chwatt (1980). Reproduced by permission of William Heinemann Ltd

breeding place for *Anopheles gambiae* or *Anopheles culicifacies* than the original swamp. An entomologist should be consulted.

The effectiveness of environmental measures varies greatly between countries and regions, being most successful in the drier areas. In Asia, where the major vector species of mosquito are short-lived and bite man relatively rarely, they are likely to give better results than in many parts of Africa where *Anopheles gambiae*, a robust long-lived species feeding predominantly on man, is the principal vector. It is such an efficient transmitter of malaria that it has to be reduced to very small numbers for control to be effective.

Malaria is not only relevant to engineers when they work on control programmes. Borrow pits, culvert silt traps, rainwater cisterns, and other bodies of water created by unthinking engineers and architects have quite often become breeding sites and been instrumental in furthering the spread of the disease. Engineering which leaves a trail of malaria behind it is bad engineering, and the control of mosquito-breeding must be considered whenever a new body of water is planned in an actual or potential malarial area (Figure 15.2). It is particularly important to leave drainage gaps in banks of spoil, and to install enough culverts in embankments to avoid the build-up of standing water behind them.

The control of malaria is especially important where a large labour force is gathered together — for a construction project, for instance — and where some of these workers are migrants from upland areas who may lack resistance to malaria. Not only do they suffer themselves, but they also increase the risk of epidemics among others by providing a focus from which infection spreads. Disastrous epidemics have occurred in this way in Southern India, Central America, the Caribbean and elsewhere.

Filariasis

The mosquito-borne forms of filariasis (Bancroftian and Malayan) are caused by nematode worms which develop in the lymphatic system and release vast numbers of tiny larvae into the blood. The worms may obstruct the lymph ducts and cause swelling of the limbs known as elephantiasis (Figure 15.3), or the scrotum (hydrocele). Infection in man is chronic and serious, and symptoms develop over many years of continual biting by mosquitoes.

Mosquito control measures are more likely to be effective against filariasis than, say, malaria because filariasis transmission can only be sustained when people are bitten many times a night. In the Solomon Islands, for instance, efforts to eradicate malaria were unsuccessful, but led to the disappearance of filariasis by reducing the numbers of mosquitoes.

There are two main kinds of mosquito-borne filariasis. Bancroftian filariasis (due to *Wuchereria bancrofti*) occurs widely in

Figure 15.3 After many years of re-infection by mosquitoes carrying filariasis, the lymphatic system becomes so obstructed that the limbs swell—a condition know as elephantiasis. This lady in India with the basket on her back has elephantiasis of both legs. Filariasis does not kill, but it disfigures and cripples
(Photo: P Sharma, WHO)

the tropics (Figure 15.4). In certain areas, particularly in the towns and cities of Asia, Egypt and coastal Brazil, it is transmitted by the night-feeding *Culex pipiens* mosquitoes (Figure 15.5). *Culex pipiens* is a group or 'complex' of closely related mosquitoes (Curtis and Feachem, 1981), which vary in such physiological characteristics as ability to hibernate or lay their first batch of eggs without taking a blood meal. The member of the *Culex pipiens* complex most widespread as a vector of Bancroftian filariasis is *Culex quinquefasciatus* which is the principal vector in the coastal cities of Brazil and of East Africa, and through much of urban Asia. However, in China and Japan the vector is *C. p. pallens*, while in Egypt it is *C. p. molestus*.

Whether or not they are vectors of filariasis in a given town, *Culex pipiens* mosquitoes are the major night-biting nuisance species in many urban areas. Community participation in the control of other mosquito species is unlikely to be sustained unless *Culex pipiens* can also be controlled, so that the mosquito population is seen to fall. On the other hand, control of the *Culex pipiens* nuisance is a popular measure even where no disease transmission is occurring. For example, in the town of Caruaru in northeast Brazil, one of the most popular measures taken by the local mayor was a high-profile mosquito control programme, although no mosquito-borne disease at all could be found there.

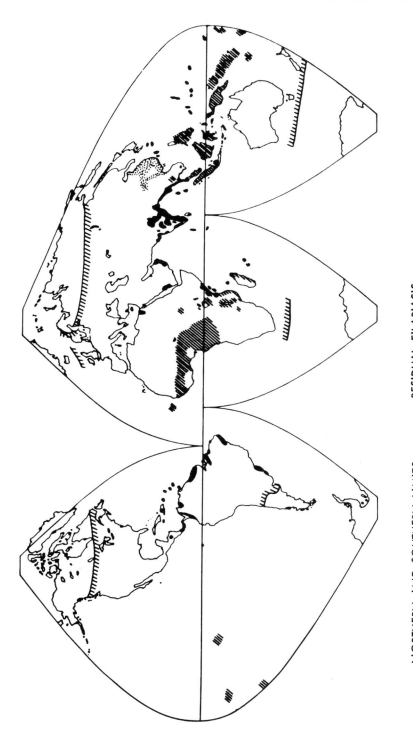

Figure 15.4 The known geographical distribution of *Culex pipiens* mosquitoes and Bancroftian filariasis
Source: From Curtis and Feachem (1981)

a

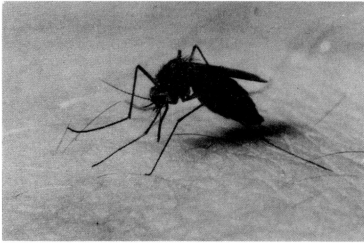

b

Figure 15.5 (a) Two larvae and one pupa of *Culex pipiens* mosquitoes suspended from the water surface. (b) An adult *Culex pipiens* taking a blood meal
(Photos: R Page)

C. quinquefasciatus breeds in dirty water and is associated with poor drainage and excreta disposal systems. Breeding in open or cracked septic tanks, flooded pit latrines, and stormwater drains is common. Insecticide control is frustrated by a readiness of this mosquito to develop resistant strains. The control of urban *C. quinquefasciatus* therefore must be through sound engineering measures for excreta, sullage, and sewage disposal, and stormwater drainage. Refuse collection must be efficient to prevent street drains becoming blocked with refuse (Figure 12.2).

In Brazil, where guppy fish eat any mosquito larvae in drainage channels, septic tanks are the main breeding site (Figure 8.13). In parts of East Africa, flooded pit latrines are responsible. Septic tanks and latrine pits can both be treated by pouring expanded polystyrene beads into them, so as to form a floating layer through which female mosquitoes cannot lay their eggs and mosquito larvae cannot breathe. These beads are often used as packing material and are very cheap. They are easier to transport when obtained from the manufacturers in pellet form; they can then be expanded on site by boiling them in water. Ideally, 2 mm diameter beads should be used, and placed to form a layer 20 mm thick (Maxwell *et al.* 1990). Experience in· Brazil, India and Tanzania has shown that the layer will remain in place for over four years.

Bancroftian filariasis is also transmitted by other mosquitoes (Figure 15.4). In much of Africa it is transmitted by the same *Anopheles* species that transmit malaria. In Polynesia, it is transmitted by day-feeding *Aedes* species and is one of the major disease problems of the area. In parts of Asia another form of filariasis (due to *Brugia malayi*) occurs and is mainly transmitted by mosquitoes of the genus *Mansonia*, although *Anopheles* spp. are also important vectors in some localities. *Mansonia* larvae breed on certain species of floating aquatic plants and can be controlled by clearing vegetation from the water. This can be done by hand, by cultivating fish (such as the grass carp) which eat the vegetation, or by making the water too saline for the plants to survive.

As with malaria, control of both mosquito-borne forms of filariasis rests on species sanitation, requiring a precise knowledge of breeding habits. Two other important types of human filariasis, onchocerciasis and loiasis, are transmitted by flies and are discussed below in Section 15.3.

Yellow fever and dengue

Aedes aegypti is a mosquito which transmits yellow fever, dengue, and some other virus infections, especially in urban areas. Many of these viruses, including yellow fever, are also present in animals such as monkeys. This makes it impossible to eradicate them solely by curing the infected people, so that control of the vector mosquito is particularly important. Dengue often appears in urban epidemics every few years. In recent years, a more virulent form known as dengue haemorrhagic fever (DHF) has become widespread. A number of insecticide campaigns have been launched against both the adults and larvae of *A. aegypti* in South-east Asia and Latin America. Insecticide resistance is widespread and some previously cleared areas have been re-invaded. *A. aegypti* is a domestic mosquito which breeds in small pools of clean water around houses — for instance in tin cans, water storage jars, rainwater tanks, flower vases, ant traps, rubber tyres, and tree holes. Its control requires community education and hygienic and intelligent behaviour around the house. Non-intermittent domestic water supply is also important, for if people have to store water in their houses, potential breeding sites are created in the storage vessels (Figure 4.2). Where they can be enforced, town by-laws may prescribe that all open vessels must be emptied every four days, and paraffin poured on any standing water which cannot be drained.

15.3 FLY-BORNE DISEASES

Four major tropical infections are carried by biting Diptera other than mosquitoes — loosely called flies in this chapter (Figure 15.6). These are African trypanosomiasis (sleeping sickness), onchocerciasis (river blindness), leishmaniasis, and loiasis (Table 15.1). In addition, flies can mechanically transmit infectious material on their legs and bodies.

Sleeping sickness

There are two forms of sleeping sickness, both transmitted by tsetse flies (*Glossina* spp.; Figure 15.6), which affect between them a huge

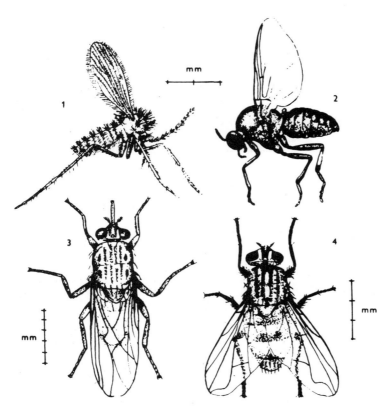

Figure 15.6 Disease-transmitting Diptera (other than mosquitoes). 1, *Phlebotomus papatasii*, a sandfly; 2, *Simulium damnosum*, a blackfly; 3, *Glossina longipennis*, a tsetse fly; 4, *Musca domestica*, the housefly
Source: From Busvine (1975). Reproduced by permission of Edward Arnold

area of tropical Africa. The first, caused by *Trypanosoma gambiense*, is transmitted in West and Central Africa by the *G. palpalis* and *G. fusca* riverine and forest-dwelling species groups. The riverine tsetse species can be considered a water-related vector (Category 4, Table 1.2), as it breeds and lives around trees and bushes along the banks of streams and around water holes, and bites people visiting the stream to water cattle or collect water for their families. Several bites by infected flies are usually required to cause sleeping sickness, and transmission is often concentrated near to tsetse breeding sites. The disease is thus potentially controlled if the need to visit breeding sites is reduced by providing adequate water supplies in the village. Tsetse control has also been achieved in parts of West Africa by clearing vegetation from the banks of streams and lakes to remove suitable breeding sites.

The other form, caused by *T. rhodesiense*, is transmitted mainly in East Africa by the *G. morsitans* group, which breeds in open woodland and feeds mainly on large game animals. Control is by separating human settlement and farming land from wild game by fencing and by sensible location of game parks. Tsetse flies do not travel far from their breeding places, and one generally applicable control measure is to clear all bush from a 200 m wide belt around human settlements.

Since the 1970s, baited traps have been used to control tsetse populations with considerable success (Laveissière *et al.* 1991). On the one hand, traps and screens have been developed which are visually attractive to the *palpalis* and *fusca* groups; on the other hand there has been the development, based in Zimbabwe, of odour-baited, pyrethroid-treated traps and screens, and pyrethroid treatment of cattle, for control of the *morsitans* group of the open savannah.

Onchocerciasis

Onchocerciasis is an infection with a filarial nematode (*Onchocerca volvulus*) causing a chronic disease which may lead to blindness. It is popularly known as river blindness and occurs in large parts of West and Central Africa and in smaller areas of South and Central America (Figure 15.7), but it is in West Africa, particularly in the Volta river basin, that the disease assumes the most terrible proportions. It has been estimated that 10% of the 10 million inhabitants of the basin (including parts of Benin, Burkina Faso, Ghana, Ivory Coast, Mali, Niger, and Togo) are infected and that 0.7%, or 70 000 people, are effectively blind.

Onchocerciasis is transmitted by the blackfly or buffalo gnat (*Simulium* spp.; Figure 15.6), which usually lays its eggs in fast flowing and well-aerated streams. The control of *Simulium* in man-made lakes and spillways is discussed in Chapter 16. Large-scale control of *Simulium* breeding is possible only by the application of larvicides to the streams and rivers of a whole river basin or area. The control of *S. neavei* in Kenya has been successfully achieved in this way but the West African vector *S. damnosum* can fly many miles in tree cover and it is very difficult to prevent recolonization from unsprayed rivers.

A massive *Simulium* control programme is now under way in 11 countries in West Africa, employing aerial spraying of all waterways with the insecticide 'Abate'. It has successfully interrupted transmission of the disease in most of the area, but the cost of this programme is extremely high and its sustainability is uncertain. Success is continually threatened by re-invasion of the area by *Simulium* from unsprayed regions and the possibility of Abate-resistant *Simulium* strains emerging. It is likely that onchocerciasis control in the remaining 23 countries, most of which are in Africa, will be based on treatment of infected people with the drug Ivermectin.

Leishmaniasis

Sandflies (Figure 15.6) transmit three tropical infections: a viral fever (papatasi fever) in the Mediterranean and Near East; a nasty bacterial fever (Oroya fever) in north-western South America, and various diseases caused by species of flagellate protozoon, *Leishmania*.

Leishmaniasis occurs in various forms throughout most areas of the tropics and subtropics (Figure 15.8). The principal distinction is between the cutaneous disease, causing sores (oriental sore) and, in Latin America, repulsive deformities of the mouth and nostrils (espundia), and the visceral disease known as kala azar, which is often fatal if not treated. Transmission is by the phlebotomine sandfly (Figure 15.6) which breeds in damp organic debris, caves, animal burrows and old anthills. In the Middle East, transmission is often concentrated around the outskirts of towns, where animals carrying the infection come in from the surrounding countryside, and where piles of refuse and builders' rubble may be important breeding sites. Breeding in pit latrines has been reported from Ethiopia. DDT house-spraying has been a successful control measure in some areas. Reduction of breeding sites in settlements is also important since the sandfly does not travel far from its birthplace.

Loiasis

Loiasis is a disease caused by infection with *Loa loa*, a filarial nematode worm, found in West and Central Africa. Endemic loiasis occurs only in or near equatorial rain forest. It is transmitted by the mangrove fly, *Chrysops* spp., which breeds in densely shaded

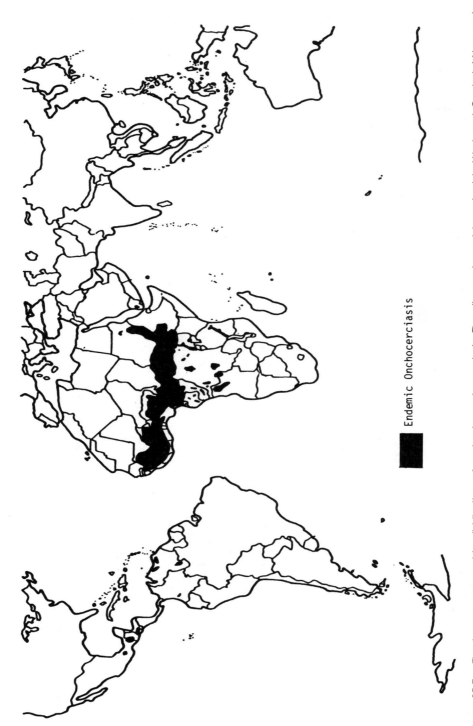

Figure 15.7 The known geographic distribution of endemic onchocerciasis. The disease is most important in West and Central Africa and especially in the Volta River basin
Source: Data from WHO

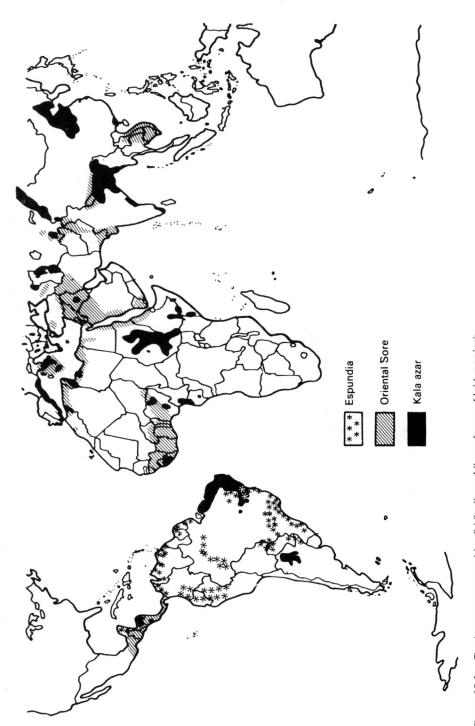

Figure 15.8 The known geographic distribution of three forms of leishmaniasis
Source: Data from WHO

mud flats or swamps near slow-moving streams. Breeding may be controlled by clearing shade vegetation and applying insecticides at breeding sites. Adult flies live in the tree tops and females descend to bite man. Clearing vegetation back at least 200 m from settlements will help to protect the community from infection.

Mechanical transmission

Flies of many species can carry faecal material in or on their bodies and thus transport pathogens contained in animal or human faeces. They provide another route for the transmission of faecal–oral infections (Table 1.2). Diarrhoeal diseases may be transmitted in this way. Flies frequently lay their eggs in excreta, and transmission is most probable when a fly visits human food having recently visited excreta. Housefly control by DDT spraying has been successful in the past but many flies are now resistant. Sanitary methods are still the best and require the removal of all breeding sites such as exposed refuse, excreta, and poorly designed latrines. A sanitary and hygienic domestic environment with adequate provision for the disposal of refuse and excreta drastically reduces fly populations.

The life cycle of the ordinary housefly, *Musca domestica* (Figure 15.6), takes about ten to fourteen days in favourable circumstances. An egg laid in organic refuse or excreta will develop into a white, maggot-like larva six to seven days after laying; the larva usually emerges and burrows into the soil for the pupal phase, emerging as a young fly seven to eleven days after laying. Three days later, the new female fly is ready to lay eggs. Refuse and excreta therefore have to remain exposed for several days for flies to breed. Collected refuse should be treated or covered as soon as possible, as it may already contain fly eggs or larvae laid before collection.

Two eye infections, conjunctivitis and trachoma, are very prevalent in the tropics and may lead to blindness. Flies are strongly implicated as contributing to the transmission of these diseases, particularly among young children, whose faces are often crawling with flies (Figure 15.9). Flies will feed on the discharge from an infected eye and transmit it to another child on their legs or in their faeces. Fly control is as outlined above and depends on good domestic hygiene and sanitation. However, conjunctivitis and trachoma transmission would continue by direct person-to-person contact and the best control method is almost certainly water supply to allow frequent bathing of infected eyes (Chapter 1). In addition, increased water availability in a dry environment means that flies have more alternative sources of moisture such as puddles, and are therefore less likely to seek it from children's eyes; it also makes it easier to keep the household cleaner, and thus less attractive to flies.

Figure 15.9 *Musca sorbens* on the face of a Masai child in Kenya. These flies may play a major role in transmitting trachoma and other eye infections (Photo: Wellcome Museum of Medical Science)

15.4 CHAGAS' DISEASE AND BUGS

Chagas' disease (or American trypanosomiasis) occurs in Central and South America and affects more than 7 million people. It is transmitted by triatomid bugs (Figure 15.10), and particularly by those species which invade human dwellings and form persistent indoor colonies. They live and breed in cracks and crannies in walls and furniture, and emerge at night once or twice a week to search for a blood meal. If the house is deserted they can survive starvation for many months.

House-spraying with residual insecticides has had some success but has led to resistant strains of bugs. Chagas' disease is a disease of the rural poor and would undoubtedly be controlled by any general improvement in rural living standards, hygiene, and rural housing (Schofield *et al.*, 1990).

15.5 LICE, FLEAS, TICKS, AND MITES

Lice, fleas, ticks and mites (Figures 15.10–12) are variously vectors of plague, relapsing fevers, several forms of typhus, and some arboviral diseases. However, with the exception of louse-borne typhus and relapsing fever, all these infections are chiefly found in animals and their vectors are not true parasites of man. They are particularly prevalent where people habitually cohabit with animals

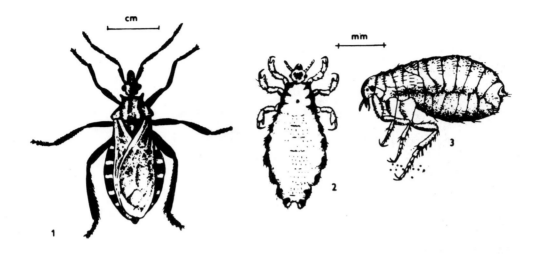

Figure 15.10 Disease vectors other than Diptera. 1, *Panstrongylus megistus*, a triatomid bug; 2, *pediculus humanus*, the human louse; 3, *Xenopsylla cheopis*, the tropical rat flea
Source: From Busvine (1975). Reproduced by permission of Edward Arnold

and poultry, and human hygiene alone cannot therefore remove the risk of infection.

Human lice (Figure 15.10) live permanently on man and do not survive starvation or changes in environmental conditions. They move from person to person at times of close bodily contact (such as sexual intercourse) or when families are huddling together in cold weather. Lice transmit epidemic or louse-borne typhus which is an infection by *Rickettsia prowazeki*. It is potentially a worldwide problem but is now mainly found in Africa and mostly at higher altitudes. It is an acute and often fatal fever. Lice also transmit louse-borne relapsing fever due to infection by a spirochaete named *Borrelia recurrentis*. It is rarely fatal.

Both louse-borne typhus and relapsing fever are human epidemic diseases which occur among people heavily infested with lice. These are usually poor people living in squalor in colder climates where they must wear plenty of clothes and sleep in crowded conditions, owing to poverty or in search of warmth. The diseases also commonly occur among soldiers. Control is effected by improving personal hygiene and thus reducing the level of lousiness. Frequent washing of the body and underclothes will virtually eliminate lice and remove the danger of an epidemic of either typhus or relapsing fever.

African tick-borne relapsing fever, spread in parts of East and Southern Africa by *Ornithodorus moubata* (Figure 15.11) a tick which hides in earth floors and walls, has been mentioned in Section 1.5. The use of beds and harder flooring material and the

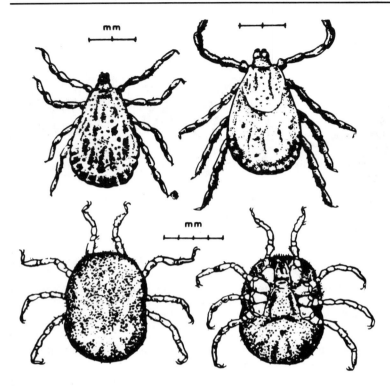

Figure 15.11 Ticks, Above, *Dermacentor andersoni*, a hard tick (left, male; right, female).
Below, *Ornithodorus moubata*, a soft tick (left, dorsal; right, ventral)
Source: From Busvine (1975). Reproduced by permission of Edward Arnold

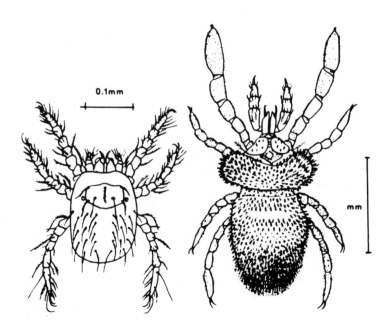

Figure 15.12 *Trombicula akamushi*, a mite vector of scrub typhus. Left, the larva; right, the adult
Source: From Busvine (1975). Reproduced by permission of Edward Arnold

smearing of cracked mud walls with wet sand should help to control this disease.

15.6 CONCLUSIONS

The control of arthropod vectors by insecticides (particularly DDT) has had some notable successes and some unfortunate failures. Environmental concern about the use of insecticides may be exaggerated and is mainly expressed in the rich and temperate countries which suffer little from the diseases discussed in this chapter. Additionally, the vast majority of insecticide application is in the agricultural sector and the volumes applied for medical reasons are relatively small.

The rising cost of DDT, and of its expensive alternatives, is a more real deterrent to its continued use in poor countries. Also of very real concern is the mounting level of resistance to insecticides in disease vectors. Many mosquitoes, flies, lice, fleas and bugs now have resistance in various parts of the world. No solution to this problem of resistance is currently in sight.

We have seen in this chapter that engineering and environmental measures are effective against a variety of vectors. Much research is going on into more sophisticated methods, such as sterile male release, genetic manipulations, predators and vector pathogens, but these have led to few practical successes so far. Engineering and environmental measures continually reappear as the long-term solution and they are particularly successful in countries where high levels of social control and community involvement are possible.

15.7 REFERENCES AND FURTHER READING

Bruce-Chwatt, L.J. (1980). *Essential Malariology* (London: Heinemann).

Busvine, J.R. (1975). *Arthropod Vectors of Disease*, The Institute of Biology's Studies in Biology No. 55 (London: Edward Arnold).

Busvine, J.R. (1980). *Insects and Hygiene*, 3rd edition (London: Chapman and Hall).

Busvine, J.R. (1982). *The Control of Domestic Flies*, Ross Bulletin No. 5 (London: London School of Hygiene and Tropical Medicine).

Cairncross, S., Rajavel, A.R., Vanamail, P., Subramaniam, S., Paily, K.P., Ramaiah, K.D., Amalraj, D., Mariappan, T. and Srinivasan, R. (1988). Engineering, mosquitoes and filariasis; a case report, *Journal of Tropical Medicine and Hygiene*, **91**, 101–106.

Curtis, C.F. (1991). *Control of Disease Vectors in the Community* (London: Wolfe Publishers).

Curtis, C.F. and Feachem, R.G. (1981). Sanitation and *Culex pipiens* mosquitoes: a brief review, *Journal of Tropical Medicine and Hygiene*, **84**, 17–25.

Davidson, G. (1988). *Insecticides*, Ross Bulletin No. 1 (London: London School of Hygiene and Tropical Medicine).

Laveissière, C., Vale, G.A. and Gouteux, J.-P. (1991). Bait methods for tsetse control, in *Control of Disease Vectors in the Community*. Curtis, C.F. (ed) (London: Wolfe Publishers).

Maxwell, C.A., Curtis, C.F., Haji, H., Kisumku, S., Thalib, A.I. and Yahya, S.A. (1990). Control of Bancroftian filariasis by integrating therapy with vector control using polystyrene beads in wet pit latrines, *Transactions of the Royal Society of Tropical Medicine and Hygiene*, **84**, 709–714.

Schofield, C.J., Briceño-Leon, R., Kolstrup, N., Webb, D.J.T. and White, G.B. (1990). The role of house design in limiting vector-borne disease, in *Control of Disease Vectors in the Community*. Curtis, C.F. (ed) (London: Wolfe Publishers).

WHO (1976a). *Resistance of Vectors and Reservoirs of Disease to Pesticides*, Technical Report Series No. 585 (Geneva: World Health Organization).

WHO (1976b). *Epidemiology of Onchocerciasis*, Technical Report Series No. 635 (Geneva: World Health Organization).

WHO (1980). *Environmental Management for Vector Control*, Technical Report Series No. 649 (Geneva: World Health Organization).

WHO (1992). *Manual on Environmental Management for Mosquito Control*. WHO Offset Publication No. 66 (Geneva: World Health Organization).

16

Dams, Irrigation and Health

16.1 GENERAL CONSIDERATIONS

Planning for Health

Typically, health considerations are incorporated into water development projects far too late and become a series of 'fire brigade operations' in which efforts are made to combat health crises which have already occurred or to head off health crises which are looming. Clearly, this is a most inefficient method and it is preferable that health planning be incorporated from the very inception of a scheme and that health considerations be borne in mind throughout the planning, construction, and operations phases. This has obvious humanitarian benefits but may well also have important economic advantages.

The area of irrigated land in the world has increased by about 1 million hectares per annum in recent decades, and special measures are required if the benefits of irrigation schemes are not to be outweighed by the damage they may do to public health. If health in the area of an irrigation scheme can be improved (or at least prevented from deteriorating), the availability and productivity of labour may well be increased. Many of the infections related to man-made lakes and irrigation are debilitating and chronic and the effect of these infections (such as schistosomiasis and onchocerciasis) on productivity may be substantial.

Efficient operation and maintenance of an irrigation scheme is a necessary condition for preventing serious health consequences. In many cases, it is also a sufficient condition, because disease vectors often thrive in overgrown vegetation, blocked channels, and fields where excess water has been wastefully applied or ineffectively removed by drainage.

It is sometimes suggested that the adverse effects of water development are mainly associated with large projects, particularly large dams. This is not necessarily true where vector-borne diseases are concerned. With schistosomiasis and guinea worm, for example, the number of potential transmission sites is more closely related to the length of a dam's shoreline than to its volume. The same is true of mosquito breeding sites, with the additional factor that the waves found on a large lake are likely to inhibit mosquito breeding along the shore. A large number of small dams can therefore create many more sites than a single large dam holding the same amount of water.

Health planning only takes place in the development of an irrigation scheme if money is spent and efforts are made. In practice, this is unlikely to occur unless health considerations are included in the Terms of Reference for the initial feasibility studies for the scheme (Tiffen, 1989).

A specific percentage of the project costs should be devoted to health planning and preventive measures. Epidemiologists, with detailed knowledge of the local disease ecology, must be employed and a process of bargaining must go on between the epidemiologists (pressing for various preventive measures) and the economists and engineers (pressing for savings in construction costs). Too often irrigation schemes are designed by teams of agronomists, engineers, and economists with little or no epidemiological input.

The construction phase

The first health consideration (chronologically) is the protection of the health of the people who will build the scheme. These people may be strangers to the area who have been brought in as labour and skilled personnel. Some of them will not have immunity to the diseases of the area and their health needs protection. Preventive measures must comprise adequate water supplies and sanitation facilities in the construction camps. Construction camps often become towns or large villages after construction has ended and this provides another incentive to invest for the future by choosing the site carefully and installing good sanitary infrastructure from the very beginning.

Appropriate immunization and prophylaxis should be given to incoming workers. Measures can be taken in the area to reduce disease vectors, and consideration must be given to the non-water-related health problems which are found in construction camps, such as injury, alcoholism, and venereal disease.

Displaced persons

If a lake is being created in an inhabited area people will be displaced. The resettlement and health of these people must be given careful consideration. Such elementary points as ensuring that they have sufficient to eat before agriculture becomes established in their new settlement are often overlooked. The villages and towns to which they are resettled may become central places in the region and, as with the construction camps, an investment in sanitary infrastructure will prove worthwhile. The sites of these places should be carefully selected to minimize contact between the inhabitants and any vectors of disease.

Among the most serious health problems of large construction schemes are 'frontier' diseases with animal reservoirs, such as yellow fever, sleeping sickness, Chagas' disease and leishmaniasis. These arise from the proximity of human habitations to undeveloped jungle, bush or desert in which the animal carriers of the diseases are found. Where these diseases are endemic, special precautions are advisable.

Agricultural workers

When construction is complete there will be an influx of people into the irrigated areas to exploit the new agricultural potential. Many of these people may be from distant or ecologically different areas and may lack immunity to local diseases. For instance, people from densely populated upland areas may move down to exploit irrigated lowland and thus be exposed to epidemics of malaria. In addition, fishermen move in to settle around the lake and they too may lack necessary immunities.

Changes in the local ecosystem which affect disease patterns

The creation of a lake and an irrigation system establishes a radically new ecological regime and with it a new pattern of infectious disease. New diseases may be introduced but, more probably, diseases already present will undergo changes in their epidemiology and prevalence. Without any health planning some infections would increase in prevalence while others might decrease. The purpose of health planning is to promote the decreases and eliminate the increases.

A guide to forecasting the likely changes in vector-borne diseases is given by Birley (1989). Surtees (1970) identified six ecological consequences of irrigation which affect mosquito populations:

(i)   simplification of habitat;
(ii)   increased surface water;
(iii) higher water table;
(iv) changes in water flow;
(v)   climatic changes towards wetter and cooler conditions; and
(vi) changes in the human population (probably increases).

Table 16.1 lists the mosquitoes which may be affected by irrigation development and relates them to their role as vectors of arboviruses, malaria, and filariasis. Figure 16.1 depicts the changes in man-biting mosquitoes that resulted from irrigation in Kenya. Of special importance is the increase in *Anopheles gambiae* which is the major African vector of malaria and also carries arboviral infections and Bancroftian filariasis.

Changes in insect and snail populations, in conjunction with other direct effects of irrigation projects, may cause marked changes in the transmission of the following infections:

- schistosomiasis;
- onchocerciasis;
- malaria;
- arboviral infections; and
- filariasis.

It is these diseases, above all others, which require special preventive action when planning an irrigation development. These will be reviewed in turn except for schistosomiasis which is treated separately in the next chapter.

16.2 ONCHOCERCIASIS

Onchocerciasis, also known as river blindness and transmitted by the blackfly *Simulium*, is discussed in Chapter 15. Irrigation development may have a marked impact on the blackfly population and control of the vector can only come from detailed knowledge of its behaviour. The larvae and pupae of *Simulium* usually develop while attached to rocks or vegetation not more than 0.15 m below the surface of a well-aerated, clear, and briskly flowing stream. The larvae of *Simulium damnosum* in West Africa are typically found in streams with velocities of 0.7–1.2 m/s. A man-made lake may eliminate such conditions by flooding out many miles of rapids above the dam site (as in the Kainji lake on the river Niger) or by increasing the suspended solids or organic loads of the stream. Below the dam site, new turbulent water areas may be created by spillway releases and these may promote *Simulium* breeding (see Section 16.6).

Simulium may be controlled chemically by introducing a suspension of DDT particles (the size of which must be very carefully chosen because their larvae are selective filter feeders) at a concentration of 0.025 mg/l. Spillways are a good site for the introduction of the insecticide because the turbulence will ensure efficient mixing. This small concentration was sufficient to remove *Simulium* from 100 km of Nile rapids.

At Kainji in Nigeria, prior to dam construction, onchocerciasis

Table 16.1 Major mosquito species favoured by dam construction and irrigation

Mosquito species	Location	Disease transmitted
Anopheles albimanus	S. America	A,M
A. barbirostris	India, Indonesia	M,F
A. coustani	Africa	—
A. darlingi	S. America	M
A. freeborni	USA	—
A. funestus	Africa	A,M,F
A. gambiae	Africa	A,M,F
A. hyrcanus	Trop. Asia	M
A. maculipennis	Europe	A,M
A. pharoensis	Africa	M
A. pseudopunctipennis	S. America	M
A. quadrimaculatus	USA	A,M
A. vagus	India	—
Aedes dorsalis	USA	A
Ae. normanenis	Australia	A
Ae. vexans	Europe, USA	A
Ae. vigilax	Australia	A
Culex annulioris	Africa	—
C. annulirostris	Australia	A
C. antennatus	Africa	A
C. bitaeniorhynchus	Asia, Australia	A
C. erraticus	USA	—
C. modestus	Europe	A
C. tarsalis	USA	A
C. territans	USA	—
C. tritaeniorhynchus	Asia	A
C. univittatus	S. Africa	A
Mansonia africana	Africa	A
M. annulata	Trop. Asia	F
M. annulifera	Trop. Asia	F
M. indiana	Trop. Asia	F
M. uniformis	Africa	A
Culiseta inornata	USA	A
C. melanura	USA	A
Psorophora confinnis	USA	A

A — arboviruses
M — malaria
F — filariasis
— not known to be a vector
Source: After Surtees (1975)

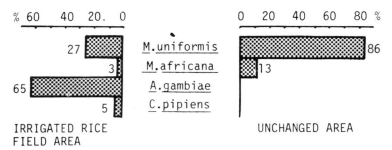

Figure 16.1 The proportions of various mosquito species biting man in an unchanged and in an irrigated area a few miles apart near Kisumu, Kenya. The increase in *A. gambiae* is especially alarming since this mosquito is the major African vector of malaria and also carries Bancroftian filariasis and arboviral infections.
Source: From Surtees *et al.* (1970). Reproduced by permission of the Royal Society of Tropical Medicine and Hygiene

was a major health problem with 5.7% of the population suffering from blindness and 49% infected. Because of this, construction camps were built about 15 km from the river. *Simulium* larvae were attacked by dosing streams with DDT for thirty minutes every ten days. These methods removed larvae from the treated stretches but adults rapidly reappeared and colonized. Since the dam was built, *Simulium* breeding has ceased upstream due to the lake and for miles downstream due to the new flow regime. This has made an increase in human settlement possible in the Kainji area.

16.3 MALARIA

The general problems surrounding *Anopheles* control, and the experience with DDT and other residual insecticides used to kill adult mosquitoes, are reviewed in Chapter 15. Control of *Anopheles* breeding is complex because measures must be tailor-made for the particular species of mosquito vector which is causing concern ('species sanitation'). The necessary knowledge cannot be simply transferred from one area to another, because different environments and species can respond quite differently to a particular intervention.

Nevertheless, a few basic measures have general application, particularly where new lakes are concerned. The first, of course, is to choose a reservoir site far from human settlements, although this is not usually possible. Where large lakes are planned, antimalarial measures should be concentrated on sections of shoreline within 2 km of villages and towns.

Another generally applicable measure, before beginning to fill the lake, is to clear vegetation from as much as possible of the area to be submerged. Where the area is too large for this, the clearing should at least include a belt extending above and below the shoreline, where mosquitoes are most likely to breed. Excavation and filling may also be used before impounding, to obtain steep banks and avoid a shallow, overgrown margin.

Mosquito breeding in fields of irrigated crops, such as rice, is exacerbated by continuous irrigation, by insufficient weeding, and by careless practices such as leaving fallow paddy fields flooded and allowing irrigation channels to become overgrown. Maintenance of channel banks to keep them steep and vegetation-free, and drainage of fallow fields and unused channels, are basic precautions.

Antimalarial precautions are particularly necessary in a newly irrigated area, where a lack of immunity among the population can permit very serious outbreaks.

16.4 ARBOVIRAL INFECTIONS

Arboviral infections are caused by numerous viruses (about 300 species of arbovirus are known) and are transmitted amongst vertebrate hosts by blood-sucking arthropods — mainly culicine mosquitoes (Figure 15.1). The word arbovirus is a contraction of arthropod-borne virus.

Infection may cause acute illness leading to immunity. It is rarely fatal except in the cases of yellow fever and haemorrhagic dengue (Chapter 15). Some forms of encephalitis may cause chronic damage to the central nervous system. Arboviral infections transmitted by mosquitoes which are likely to be favoured by irrigation and dam developments are listed in Table 16.2. Control is exactly as for malaria with an understanding of vector biology being the key to species sanitation.

16.5 BANCROFTIAN AND MALAYAN FILARIASIS

These two main types of mosquito-borne filariasis have been discussed in Chapter 15. *Bancroftian filariasis* is related to any irrigation or lake development because the vectors are all water breeders. As with malaria, water development projects may increase or decrease the number of suitable breeding habitats, and the size and density of the human population, and so may affect the local epidemiology of filariasis. In the absence of control measures, suitable habitats and human population sizes and densities are likely to increase, so that filariasis may become more common and more serious.

The major vector species in many areas, *Culex quinquefasciatus*, is one of the *Culex pipiens* complex and breeds in polluted waters in and around human settlements (Figures 15.4 and 15.5). It is likely to become established in construction camps and resettlement areas. The important African vector, *Anopheles gambiae*, may well find new breeding sites associated with the lake and irrigation water and may thus pose the threat of an increase in filarial, malarial, and arboviral transmission. It is notable that rice irrigation in western Kenya caused an increase in both *Culex quinquefasciatus* and *Anopheles gambiae* (Figure 16.1). Control is based on a detailed

Table 16.2   Arboviruses transmitted by mosquitoes favoured by dam construction and irrigation

Arbovirus	Where found	Vector species
Buramba	Africa	*Anopheles funestus*
Cache Valley	North America	*Culiseta inornata*
California	USA, Canada	*Culex tarsalis*
Calovo	Central Europe	*Anopheles maculipennis*
Chikungunya	Africa, SE Asia, Philippines	*Anopheles gambiae, A. funestus Mansonia africanus, M. uniformis, Culex tritaeniorhynchus*
Ilheus	South and Central America	*Psorophora confinnis*
Japanese encephalitis	Asia, Pacific Islands	*Culex tritaeniorhynchus*
Murray Valley	Australia, New Guinea	*C. annulirostris, C. bitaeniorhynchus*
O'nyong-nyong	Africa	*Anopheles gambiae, A. funestus*
Ross River	Australia, Pacific Islands	*Culex annulirostris*
St. Louis	Americas, Jamaica	*Aedes dorsalis, Culex tarsalis*
Sindbis	Africa, India, SE Asia, Philippines	*Culex antennatus, C. univittatus*
Spondweni	Africa	*Mansonia africana, M. uniformis*
Tahyna	Europe	*Aedes vexans, Culex modestus*
Tensaw	USA	*Anopheles quadrimaculatus, Psorophora confinnis*
Tlacotalpan	Central America	*Anopheles albimanus*
Wesselsbron	Africa, Asia	*Mansonia uniformis*
Western equine enchephalitis	USA	*Culex tarsalis Culiseta melanura*
West Nile	Africa, India, Middle East, Europe	*Culex antennatus, C. univittatus C. modestus*

Source: After Surtees (1975)

knowledge of vector biology and ecology so that breeding sites may be removed, or treated with insecticides.

Malayan filariasis is transmitted mainly by *Mansonia* spp., the larvae of which attach to the roots and leaves of aquatic plants. The danger is therefore that man-made lakes may, due to eutrophication, become colonized with water hyacinth (*Eichhornia crassipes*), water fern (*Salvinia auriculata*), or water lettuce (*Pistia stratiotes*) all of which may provide suitable breeding habitats for *Mansonia*.

16.6 SPILLWAYS AND RESERVOIR RELEASE POLICIES

The design of spillways and reservoir release policies may affect fly, mosquito, and snail populations and thus diseases which are related to flies, mosquitoes, and snails. In particular, the fluctuation of water levels and the release of sudden flushes of water may prevent colonization by snails and mosquito larvae. Three main diseases are affected by spillway design and release policy: malaria, onchocerciasis and schistosomiasis. Measures designed to reduce one of these infections may promote another. Therefore, when all three are present or in danger of being introduced, each of these diseases must be considered in detail before decisions are taken.

For instance, in areas such as West Africa where *Anopheles gambiae* is the main vector of malaria, a policy of level fluctuation and flushing might well reduce the local prevalence of onchocerciasis and schistosomiasis (by preventing *Simulium* breeding and reducing snail populations) but at the same time malaria transmission could be increased due to increasing numbers of *A. gambiae* which can breed in transient sunny puddles such as those created by fluctuating water levels. The same applies to *A. punctulatus*, an important vector in the Pacific, and to *A. culicifacies* in India (Table 15.2).

However, this example also illustrates the way different species will respond differently to the same intervention, because water level fluctuation was used successfully in the Tennessee Valley in the USA to control another malaria vector, *A. quadrimaculatus*. The shoreline was cut to give a 1:8 slope, and spillway discharges were regulated so that water levels rose and fell each week by about 0.3 m, and thus prevented the establishment of the static, shallow, shaded water this mosquito requires to breed. Clearly, then, it is important to identify the main vectors in the area and their habits before prescribing a reservoir release policy; but some general comments can be made.

Onchocerciasis and reservoir release

The pattern of release of water from large man-made lakes tends to depend on the ratio of lake volume to the average flow of the dammed river. If the lake is small in relation to the mean annual river flow (as at Kainji in Nigeria), spillway release needs to be frequently adjusted, leading to fluctuating water levels downstream of the dam. This will discourage *Simulium*. However, if the lake is large (as at the Volta Dam), spillway discharge may be fairly constant and favourable breeding conditions may then be established downstream. Breeding in the spillway itself is also common and siphon spillways (see below), or rapidly fluctuating discharges, are possible control devices. However, *Simulium* development is rapid — sometimes as rapid as five days — so that any level fluctuations or intermittent discharges must be maintained with a high frequency if *Simulium* breeding is to be prevented.

In numerous small rural dams in Burkino Faso a variety of spillway designs has provided good breeding sites for *Simulium*. The construction of these dams has spread onchocerciasis transmission northwards and away from major river valleys into areas previously protected by their distance from good breeding sites. Maximum

Simulium breeding occurs between the end of the rains and the period of no flow (i.e. between September and December in Burkino Faso) when water levels and spillway discharges are falling.

Different *Simulium* species will colonize different spillways, depending largely on the speed of flow of the water. Various design modifications have been tried in Burkina Faso, but these failed to prevent *Simulium* breeding. The only spillways guaranteed to prevent *Simulium* breeding are sluice gates and siphons, with which the spillway may be completely dried out periodically. However, these are expensive, have operation and maintenance problems, and require an attendant.

Siphon spillways

Siphon spillways (Figure 16.2) cause the continuous fluctuation of water levels in the dam and intermittent discharges downstream. The water level will rise, while the siphon primes, and there will be no discharge. Then, when priming is complete, the water level will fall rapidly and there will be a sudden discharge of water downstream. Such a regime has been recommended for the simultaneous control of *Simulium* breeding, *Anopheles* breeding, and snails. The spillway is designed to give the desired flushing frequency on the basis of the expected flow during the season in which the propagation of the vector is at its peak. This is not necessarily the season of peak rainfall, and tends to be slightly later.

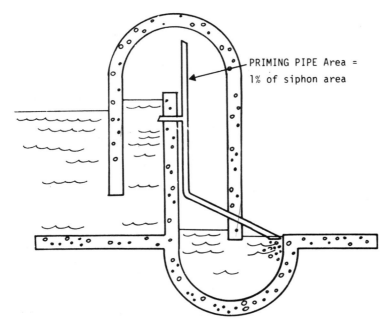

PRIMING PIPE Area = 1% of siphon area

Figure 16.2 A siphon spillway. Water flowing down through the priming pipe sucks air with it from the dome, so that the chamber fills with water and siphoning begins *Source*: From Krusé and Lesaca (1955). Reproduced by permission of Johns Hopkins Press

Simulium will not be able to breed in the spillway itself when a siphon is used. *Anopheles* control is more problematical (see above), and a detailed knowledge of the local anopheline vector ecology is required. Downstream of the dam, the flushing action of the periodic discharges will have some effect on anopheline breeding. Work in the Philippines has shown that flushing controls the larvae of *A. minimus* from 200 m to 1500 m downstream of a siphon spillway. Flushing not only washes out larvae but also changes stream ecology and removes much vegetation.

It may not be feasible to control the snail hosts of schistosomiasis by fluctuating water levels in the dam (Chapter 17). Downstream the periodic discharges, and associated vegetation changes, will certainly discourage the snail population, but this will affect only a limited length of stream below the dam.

Siphon spillways have a number of practical disadvantages. They may well overtop in periods of high runoff, so that additional spillways must be provided. Siphons in the Philippines have suffered greatly from vandalism and require good maintenance and supervision. During the filling phase there will be no flow downstream and therefore downstream water-users may be annoyed and try to create a permanent flow by breaking the siphon. Dams are always created as a water conservation device and a siphon spillway determines a specific release policy which may well not be that which would be selected from a water resources standpoint. In any case, during periods of the year when inflow to the reservoir is less than evaporation and losses, water levels will steadily fall and the siphon will be inoperative. They are thus inappropriate for small impoundments in regions having a pronounced dry season, and this includes a large part of the tropics and subtropics.

16.7 REFERENCES AND FURTHER READING

Birley, M.H. (1989). *Guidelines for Forecasting the Vector-Borne Disease Implications of Water Resources Development*, PEEM Guidelines Series No. 2 (Geneva: World Health Organization).

Bradley, D.J. (1977). The health implications of irrigation schemes and man-made lakes in tropical environments, in *Water, Wastes and Health in Hot Climates*, Feachem, R., McGarry, M. and Mara, D. (eds) (London: John Wiley). pp. 18–29.

Burton, G.J. and McRae, T.M. (1965). Dam spillway breeding of *Simulium damnosum* (Theobald) in Northern Ghana. *Annals of Tropical Medicine and Parasitology*, **59**, 405–411.

IRRI (1988). *Vector-borne Disease Control through Rice Agroecosystem Management* (Los Baños, Philippines: International Rice Research Institute).

Krusé, C.W. and Lesaca, R.M. (1955). Automatic siphon for the control of *Anopheles minimus* var. *flavirostris* in the Philippines, *American Journal of Hygiene*, **61**, 349–361.

Mather, T.H. and That, T.T. (1984). *Environmental Management for Vector*

Control in Rice Fields, FAO Irrigation and Drainage Paper No. 41 (Rome: Food and Agriculture Organization).

Quélennec, G. and Simonkovich, E., and Ovazza, M. (1968). Recherche d'un type de déversoir de barrage défavorable à l'implantation de *Simulium damnosum* (Diptera, Simulidae), *Bulletin of the World Health Organization*, **38**, 943–956.

Rydzewski, J.R. (1987). *Irrigation Development planning; an Introduction for Engineers* (London: John Wiley).

Stanley, N.F., and Alpers, M.P. (1975). *Man-made Lakes and Human Health* (London: Academic Press).

Surtees, G. (1970). Effects of irrigation on mosquito populations and mosquito-borne diseases in man, with particular reference to rice-field extension, *International Journal of Environmental Studies*, **1**, 35–42.

Surtees, G. (1975). Mosquitoes, arboviruses and vertebrates, in *Man-Made Lakes and Human Health*, Stanley, N.F. and Alpers, M.P. (eds) (London: Academic Press) pp. 21-34.

Surtees, G., Simpson, D.I.H., Bowen, E.T.W. and Grainger, W.E. (1970). Ricefield development and arbovirus epidemiology, Kano Plain, Kenya, *Transactions of the Royal Society of Tropical Medicine and Hygiene*, **64**, 511–518.

Tiffen, M. (1989). *Guidelines for the Incorporation of Health Safeguards into Irrigation Projects through Intersectoral Cooperation*, PEEM Guidelines Series No. 1 (Geneva: World Health Organization).

WHO (1980). *Environmental Management for Vector Control*, Technical Report Series No 649 (Geneva: World Health Organization).

17

Schistosomiasis

17.1 INTRODUCTION

Schistosomiasis, also known as bilharzia, is an important disease whose vector is not an arthropod, but an aquatic snail. Different strategies for its environmental control are therefore appropriate and these are considered in this chapter.

Human schistosomiasis is mainly caused by one of four species of trematode worms:

> *Schistosoma japonicum* — found in East Asia and the Philippines and infecting domestic and wild animals as well as man.
> *Schistosoma mansoni* — found in Africa, the Middle East, South America, and the Caribbean and infecting man and some animals.
> *Schistosoma haematobium* — found in Africa and the Middle East and rarely infecting animals.
> *Schistosoma intercalatum* — found in Cameroon, Congo, Gabon, and Zaire.

Figures 17.1 and 17.2 show the geographical distribution of the three main species. *S. intercalatum* is of relatively minor importance.

People who are infected excrete schistosome eggs in their faeces (*S. japonicum, S. mansoni, and S. intercalatum*) or urine (*S. haematobium*). These eggs must find water, where they will hatch into miracidia. These tiny swimming organisms will locate and infect an aquatic snail of a particular species as follows:

> *Schistosoma japonicum* infects snails of the genera *Oncomelania* and *Tricula*.
> *Schistosoma mansoni* mainly infects snails of the genus *Biomphalaria*.
> *Schistosoma haematobium* and *S. intercalatum* mainly infect snails of the genus *Bulinus*.

The individual snail species prefer different habitats but, very roughly speaking, *Bulinus* species prefer still or very slowly moving

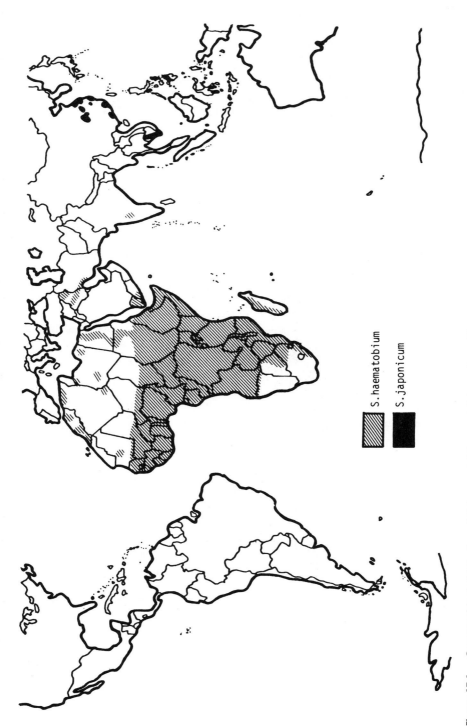

Figure 17.1 Geographical distribution of *S. haematobium* and *S. japonicum*. The areas marked are those in which transmission of these schisto-somes may be occurring. *S. haematobium* transmission is most unlikely at altitudes above 1500 m

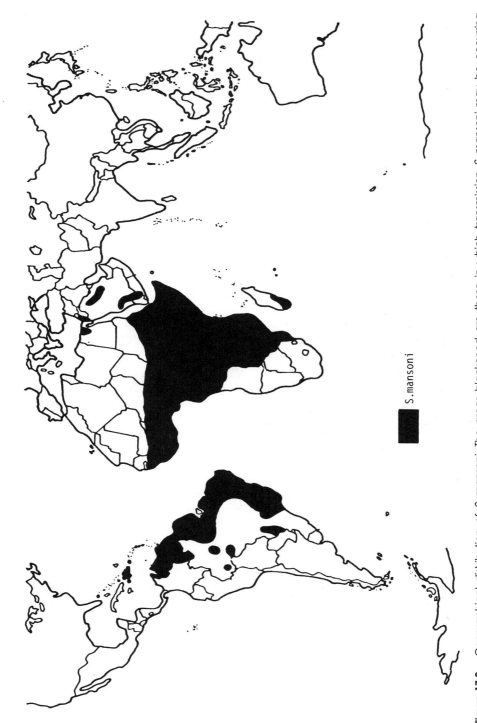

Figure 17.2 Geographical distribution of *S. mansoni*. The areas blackened are those in which transmission *S. mansoni* may be occurring. *S. mansoni* transmission is most unlikely at altitudes above 2000 m

S.mansoni

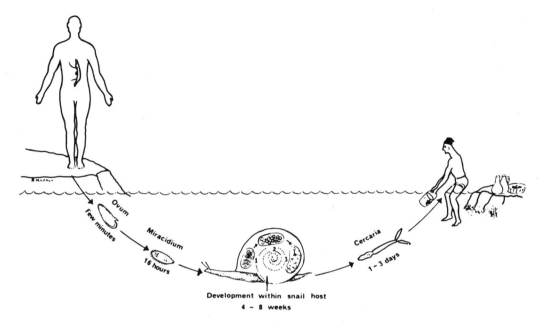

Figure 17.3   The life cycle of the schistosome worms which infect man
Source: From Jeffrey and Leach (1975). Reproduced by permission of Churchill Livingstone

water and are often found in small pools and water holes, whereas
Biomphalaria can live in gently flowing water and tend to occur
in streams and irrigation systems. *Oncomelania*, on the other hand,
are amphibious and have a protective door to their shell to help
them resist drying. Even without this door, most of the other vector
species have a pronounced capacity to survive dry periods, usually
burrowing into mud.

Some time after infection the snail will shed many *cercariae*,
a swimming form of the schistosome, which will infect their next
human host through the skin when it is immersed in water or, less
commonly, when a person drinks infected water. The life cycle is
shown diagrammatically in Figure 17.3.

The prevalence of the infection is greatest among children and
young people in the 10–20 years age group due to the high frequency
of their contact with snail-infested water. The majority of persons
infected are apparently without symptoms. Symptoms arise, not from
the adult worm, but from the eggs which fail to escape in the urine or
faeces. These symptoms increase gradually and severity is related
to worm load. General lassitude and debility are often reported.
A proportion of those infected develop grave long-term effects
which include serious damage to the kidneys, liver and urinary
tract. A common symptom of *S. haematobium* schistosomiasis is
the presence of blood in the urine.

Perhaps 300 million people are infected worldwide. Schistosomiasis is endemic in 71 countries having a total population of at least 1500 million people. It occurs in almost every country in Africa and has become a major problem around most man-made lakes and irrigation areas. Control measures have usually met with only moderate success and the disease is becoming more prevalent in many regions. The limited success achieved in combating schistosomiasis once it has become established in an area provides a strong incentive for preventive action to be taken well before any serious problem has occurred.

Control measures may be grouped in the following way:

(i)   treatment by drugs (chemotherapy) of infected persons to reduce the number of viable eggs being released into the environment;

(ii)   reduction in the snail population by chemical molluscicides;

(iii)   reduction in the snail population by engineering means;

(iv)   reduction in the release of schistosome eggs into the environment by engineering means;

(v)   reduction in the need for contact with infected bodies of water by engineering means.

The role of mass chemotherapy in schistosomiasis control is increasing due to the development of greatly improved drugs. In particular, oxamniquine (against *S. mansoni*) and praziquantel (against all three major schistosome species) are key weapons in major control programmes. Nevertheless, it is acknowledged that engineering and environmental measures have an important role to play alongside drug therapy in cutting the cost and increasing the effectiveness of disease control. Snail control and chemotherapy ((i)–(iii) above) have impact on schistosomiasis only, whereas methods based on sanitation (iv), and improved water supplies (v) will have multiple health benefits.

Environmental control measures frequently require some change in human behaviour if they are to be effective. Thus, water supplies reduce contact with infected water only if they are used, and it is useful to build irrigation canals far from human settlements only if new settlements are not allowed to spring up beside the canals. Health education on its own will not necessarily lead to these changes in behaviour, and thought must be given to the additional means to be used. Legislation may be needed.

There remains considerable debate concerning the most cost-effective mix of the possible methods listed above. Recent work has indicated that a combination of chemotherapy, mollusciciding, and engineering measures are required and that at least some of these measures must be maintained if disease levels are not to rise back to, or above, their initial values.

17.2 ENGINEERING METHODS IN SCHISTOSO-MIASIS CONTROL

Engineering methods of schistosomiasis control may be classified as follows:

(i)   hygienic disposal of faeces and urine to prevent the release of eggs into the environment;

(ii)   improved water supplies, washing facilities, and other measures to reduce the need for contact with polluted water;

(iii)   appropriate channel design;

(iv)   adequate drainage;

(v)   improved irrigation practices;

(vi)   barriers to prevent snail drifting;

(vii)   fluctuation of water levels;

(viii) measures to facilitate mollusciciding;

(ix)   miscellaneous other measures.

These will be considered in turn. The recommendations made refer primarily to the control of *S. mansoni* and *S. haematobium*. Control of *S. japonicum* is more difficult because the snail host, *Oncomelania,* is especially resistant to environmental modification and to mollusciciding, and the substantial non-human reservoir (cows, buffaloes, dogs, pigs, cats, rodents, and others) reduces the effectiveness of sanitation and chemotherapy.

Excreta disposal

If all human wastes were disposed of hygienically and underground the transmission of human schistosomiasis would cease immediately. Eggs must reach fresh water within three weeks of excretion to survive, and may not even survive until then. Once in water they may not hatch for several days and thus may be carried considerable distances with the water flow. After hatching, the free-living larval stage (miracidium) must reach a suitable snail host within twenty-four hours to avoid death (Figure 17.3). This part of the life cycle, between the human host and the snail, can therefore be broken if suitable toilets are used. In addition, many other infections would be greatly reduced by improved excreta disposal.

However, a snail infected by a single miracidium can produce many thousands of cercariae over the following months. Contamination of a water body by a single infected person can therefore transmit schistosomiasis to a whole community. Thus sanitation will only prevent schistosomiasis transmission if toilets are used by the whole community at all times; in practice, this is very difficult to achieve. In areas where *S. haematobium* is prevalent (Figure 17.1), these problems are compounded by the need to dispose hygienically of urine as well as faeces. In addition, latrines must be used in the fields, and on the way to the fields, and not just in the village or camp.

Field studies on the role of sanitation in schistosomiasis control have for the most part been poorly designed and inconclusive. Studies in Egypt showed that sanitation was ineffective because it was applied only in the home and not in the fields. Ensuring that villages and work camps are not located near reservoirs, lakes, or major canals is also important in reducing the number of schistosome eggs in faeces and urine which reach snail-infested waters.

Reduction of contact

After four to eight weeks in the snail, the second free-living stage (cercariae) are shed into the water (Figure 17.3) where they must find a suitable host within forty-eight hours. A second control method is therefore to provide adequate domestic water supplies and washing facilities which will reduce the need for people to be in contact with infected water (Figure 17.4). These facilities also have an impact on many other infections. However, they must be at least as convenient as the existing bodies of surface water if they are to be used in preference. For instance, a simple tap or pump in the village is not so convenient for washing clothes as a stream, for which no tub is required and where the natural flow provides a rinsing action (Figure 17.4).

The new facilities must provide for the various needs which bring the population into contact with infected water. For example, the highest prevalence of schistosomiasis is usually among children, much of whose contact with surface water is for play (Figure 17.4). One way to meet their needs is to provide special pools for them, created from old borrow pits or dug where the water table is high, kept clear of snail-harbouring vegetation, and separated from any adjacent potentially infected bodies of water. Without rigorous maintenance of the pools, however, they become major schistosomiasis transmission sites.

Recent studies indicate that a fairly high level of water supply provision, such as house connections for most families, may be necessary before a substantial reduction in water contact, and thus schistosomiasis transmission, is achieved. Water supplies are less effective in controlling the disease amongst certain agricultural workers and fishermen who must work in water and therefore expose themselves to risk.

Various other measures may be employed to reduce human contact with infected water, such as building bridges over channels or streams to obviate wading and piers for boarding boats, and locating communities away from bodies of water. The latter is especially valuable, but it is not enough to leave a symbolic few

Figure 17.4   Schistosomiasis transmission on Leyte, Philippines. The children may become infected while swimming and their mother may become infected when she rinses the clothes. The children are clearly at much greater risk than their mother and will typically be more heavily infected. The slightly swollen abdomen of the child on the bank is suggestive of schistosomiasis (Photo: WHO)

hundred yards between the edge of a village and the nearest canal, as has sometimes been the practice. The minimum distance to be effective is likely to be 2 km or more.

Appropriate channel design

Irrigation water may be transported in covered channels or pipes with benefits to water conservation as well as water quality and snail control. However, the cost of this is usually so high that open channels are used.

With a certain combination of water velocity, channel lining, and vegetation regime, snails will be unable to colonize a stretch of water channel ('channel' here referring to either the water distribution system or the drainage system). In general, the less the vegetation and the higher the velocity, the lower the probability of snail colonization. Channel lining helps to reduce vegetation and to increase velocity as well as reducing maintenance costs, seepage, and erosion. The velocity of flow in a channel may be computed from Manning's formula:

$$V = \frac{R^{2/3} S^{1/2}}{n}$$

where V = mean velocity (m/s);
$\quad\quad R$ = hydraulic radius (m) — cross-sectional water area divided by wetted perimeter;
$\quad\quad S$ = longitudinal slope;
$\quad\quad n$ = a coefficient indicating surface roughness, known as Manning's n.

Since snails live on the bottoms or sides of channels, it is necessary to know the velocity at these points — the 'peripheral velocity' — which may be estimated as $0.55V$.

The hydraulic radius is principally a function of the size of the channel. The longitudinal slope along the channel is hard to increase as channels must follow the ground surface. However, where natural streams are concerned, the slope may be increased by straightening them.

The values of Manning's n and the relative costs of various linings are given in Table 17.1. The choice of lining can substantially affect n and therefore V for given values of R and S. In assessing the cost of lining in relation to the benefits, the main benefits are never in disease control but always in the reduction of maintenance cost, and sometimes in the control of erosion and seepage (a quarter to a half of the water fed into an irrigation system may be lost by seepage). Snail control will be a side-benefit from lining.

It has been generally observed that snails are seldom found in channels having a mean velocity greater than 0.30–0.35 m/s. Experimental work on smooth plastic channels showed that mean velocities of the order of 0.6–0.8 m/s were necessary to dislodge snails. These correspond to peripheral velocities of 0.3–0.4 m/s and are considerably higher than field observations. This information suggests that channels kept clear of vegetation, and with mean velocities of over 0.6 m/s, have a good chance of being snail-free.

However, velocities of over 0.6 m/s are fast enough to cause serious erosion to some unlined channels. Table 17.2 shows the estimated maximum permissible velocities for different types of channel bed. It is clear that, if velocities are to be kept up, channels need lining in some circumstances.

Drainage

The drainage or filling-in of natural shallow pools and seepages helps to reduce the number of snail habitats. In irrigation schemes,

Table 17.1 Relative costs and values of Mannning's *n* for various channel linings

Lining	Relative total cost per unit length	Manning's *n*
Concrete,75 mm unreinforced	6.0	0.014
Concrete,50 mm unreinforced	5.0	0.014
Asphaltic concrete, 50 mm	5.0	0.014
Asphaltic membrane, buried	2.2	0.025
Prefabricated plank-type, exposed	4.3	0.015
Asphalt-coated jute, exposed	2.3	0.016
Asphalt-coated jute, buried	3.5	0.025
Polyethylene film, 10 mm, buried	1.7	0.025
Vinyl film, 10 mm, buried	2.1	0.025
Butyl sheeting, 32 mm, buried	3.7	0.025
Butyl sheeting, 60 mm, exposed	5.0	0.012
Earth compacted, 1 m	2.2	0.025
Earth compacted, 0.3 m	1.0	0.025
Natural stream, clean and straight	—	0.025–0.03
Natural stream, weedy and winding	—	0.05–0.15

Source: After McJunkin (1970)

Table 17.2 Maximum velocities in metres per second in earth canals

Nature of canal bed	Discharge in cubic metres per second								
	0.5	1.0	2.0	3.0	4.0	10.0	15.0	20.0	30.0
Silt, fine sand, light sandy loam	0.37	0.39	0.41	0.43	0.45	0.47	0.49	0.50	0.52
Medium sandy ground	0.46	0.49	0.52	0.54	0.56	0.59	0.61	0.63	0.65
Light loam	0.53	0.56	0.59	0.61	0.64	0.68	0.70	0.72	0.74
Medium loam, medium loess, coarse sand	0.59	0.63	0.67	0.69	0.72	0.75	0.79	0.81	0.84
Heavy loam, light clay, close grain loess, very coarse sand	0.67	0.71	0.75	0.78	0.81	0.86	0.89	0.90	0.94
Fine shingle or gravel	0.73	0.77	0.82	0.84	0.88	0.93	0.96	0.98	1.02
Thick medium clay, medium gravel	0.82	0.87	0.92	0.95	0.99	1.05	1.09	1.11	1.16
Heavy clay (tertiary), coarse shingle or gravel	1.26	1.34	1.42	1.47	1.53	1.62	1.68	1.72	1.79

Source: From Kraatz (1971), quoted in McJunkin (1975). Reproduced by permission of the Food and Agriculture Organization of the United Nations

drainage has similar effects, as well as often increasing crop yields by maintaining the right level of water table. In Egypt, agricultural production has been increased by 12–25% in areas successfully drained. Snail populations usually colonize drainage systems to a greater extent than water delivery systems, so it is worthwhile to line the invert to concentrate small flows and increase water velocity. Drainage by underground pipes prevents snail growth but is more expensive. However, the cost may be partly offset by the lower maintenance requirement of underground drains and by the saving in land.

Irrigation practices

The need for drainage may be reduced by controlling unnecessary seepage, leakage, and waste of water. Farmers could be charged a higher rate for water used in excess of the amount necessary for the area they irrigate.

In general, efficient irrigation management is good for snail control. Regular maintenance of irrigation channels to prevent blockage by vegetation also helps to control snails. Typical manpower requirements for satisfactory channel maintenance are given in Table 17.3.

Intermittent irrigation can also be remarkably effective in reducing snail habitats. In one scheme in Iraq in which water was alternately supplied and shut off every five days, no snails at all were found in the irrigated area. In parts of the Philippines, where rice was formerly grown by ponding, the number of snails was reduced from 200 to less than 1 per square metre and the rice yield increased by over 50% as a result of intermittent irrigation. The practice of growing rice in standing water arose from the desire of farmers to store water in case of drought, but is unnecessary when there is a reliable irrigation system. In the Philippine case mentioned, water was applied about every ten days whenever the soil began to show signs of drying.

Not only is it necessary to allow fields to dry out periodically. This should also be done with channels, as long as it can be done without leaving small puddles in the channel bed to become mosquito breeding sites. Traditional in-field structures such as inverted syphons, pump sumps, long weirs (or duck-bill weirs) and drop structures with sunken stilling basins are frequently colonized by snails, because they retard the flow and prevent complete drying of the channel. Alternative free-draining structures are needed wherever possible (Figures 17.5 and 17.6).

The use of small reservoirs for the overnight accumulation of

Table 17.3 Approximate manpower required for the maintenance of unlined canals

Canal type	Canal capacity	Manpower required
Irrigation canals	< 1 m³/s	1 per 2000 m
	1–10 m³/s	1 per 1500 m
	> 10 m³/s	1 per 1000 m
Drainage canals		1 per 3000 m

(a)

(b).

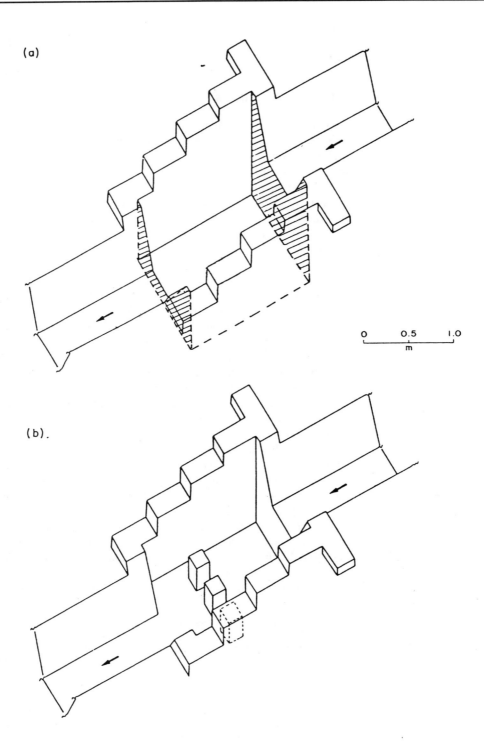

water is not recommended, because they tend to act as nurseries, producing large numbers of snails and cercariae which are then spread throughout the area. It is preferable to irrigate twenty-four hours a day or to shut off the flow upstream at night, although this may require more careful operation of the system. Where night storage is unavoidable, it should be concentrated in a few large reservoirs with steep sides, which should be completely emptied each day if possible.

The breeding of snails and release from them of cercariae is sometimes markedly seasonal, and where the seasonal pattern is known it may be possible to schedule planting, irrigation, and harvesting in such a way as to control snail populations.

Barriers

Snails often breed in an irrigation water reservoir, or become trapped there after drifting from upstream. If this is occurring, barriers may be placed on the irrigation channels leading from it. This is a measure used on the Gezira scheme in the Sudan.

Studies have indicated that drifting snails in irrigation channels mostly travel very near the surface and nearer the banks than the centre. Therefore, mechanical barriers which extend down 0.5 m from the surface, and are especially designed to trap snails near the banks, may prevent snails from carrying infection downstream. The mesh should not be coarser than 3 mm in order to catch juvenile snails. However, barriers will not prevent eventual infestation by the snails, which can be transported on birds' legs and in other ways, and so they must be used in conjunction with other control measures.

Level fluctuations

Much attention has been focused on the method of rapidly fluctuating water levels in order to strand snails on the banks where they may die through desiccation or be subsequently flushed downstream as water levels suddenly rise. Such measures may be applied to channels or lakes or to both together.

In unlined canals this is unlikely to be successful because snails can aestivate for long periods in mud and amongst vegetation. This survival is promoted by higher water tables which are a feature of irrigation areas and which encourage cool and damp conditions in the beds of dry channels. Lined canals are more suitable for control

Figure 17.5   Drop structures for lined irrigation canals. (a) Conventional design with sump. (b) Modified free-draining design as used in Zimbabwe
Source: Bolton (1988)

(a)

(b)

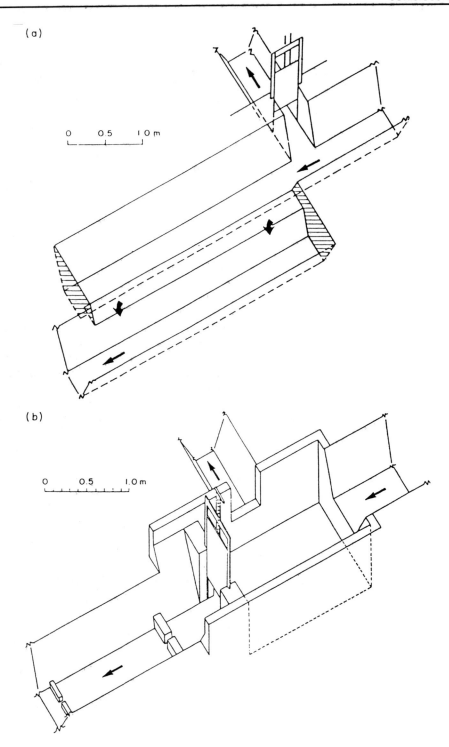

through desiccation and snails may die rapidly when exposed to sunlight on a concrete surface. Much more information is needed on the reaction of different species to this treatment and on the frequency and length of no-flow periods necessary to achieve snail elimination. Nevertheless, it is useful to design irrigation systems in such a way as to make these measures easier, by making the canals as narrow as possible and by fitting sluice gates at the tail ends.

Fluctuating water levels in a lake may also help to kill snails other than *Oncomelania*, as well as controlling the breeding of some mosquitoes. Field studies in Puerto Rico showed that drawdowns by siphon spillways (Chapter 16) of 0.5 m in a day every five to twenty days during the snail-breeding season were sufficient to remove *Biomphalaria* snails from two ponds. It appears that the method works not by killing the snails, but by preventing their reproduction. The adults could not lay eggs while they were stranded, and the few eggs produced were extremely susceptible to drying.

The rate of drawdown required depends on the beach slope, the snail species, and other factors, but a typical rate for a 1:13 slope would be about 0.3 m/h. Drawdown rates of this magnitude limit the technique to small reservoirs and small reservoirs generally have earth dams. Spillways and gates are often of prohibitive cost and therefore plastic pipe siphons have been used. However, the long stretch of horizontal pipe across the dam requires a large volume of water to prime it and, in addition, a siphon spillway on a small rural dam is generally inappropriate from an operation and maintenance viewpoint. In any case, siphon spillways cannot work at periods of the year when inflow to the dam is less than evaporation so that water levels are slowly falling (see Chapter 16).

Of all engineering measures for the control of snail populations, level fluctuation is probably the most problematical and least successful. The ultimate effect on snails is hard to predict and its effect on mosquito breeding may even be to increase populations of certain vector species (see Section 16.6). The flow regulations required are complex, prone to breakdown, and wasteful of water.

Assistance to chemical control of snails

Practically any engineering measure which helps to reduce the extent of snail habitats will also make chemical control of snails easier, more effective and less frequently necessary (Figure 17.7). Specifically, the amount of molluscicide required, and hence the cost, is proportional to the volume of water to be treated. There is therefore

Figure 17.6 Off-take structures for lined irrigation canals. (a) Conventional long-weir with gated off-take. (b) Free-draining control gate with off-take weir
Source: Bolton (1988)

an additional advantage in reducing the volume of water held in channels and reservoirs, besides speeding up the flow and keeping to a minimum the potential breeding places created. Extra-large canals which serve as longitudinal storage reservoirs require very large quantities of molluscicide. Moreover, the flow in them is so slow that molluscicides do not mix easily with the water.

Miscellaneous measures

Irrigation systems require periodic maintenance, to clear vegetation and accumulated silt, and to prevent seepage and leakage of water. This also helps to control snail infestation. One measure, which has been exploited to great effect in China, has been the burial of snails, by digging out one side of a channel and piling the soil on the other, or by filling it in and diverting its course completely. This method, of course, requires an enormous input of labour, and has to be applied repeatedly unless measures are taken to prevent recolonization by snails drifting from elsewhere.

Figure 17.7 Molluscicide application on an irrigation scheme in Egypt. Treatment of many miles of poorly maintained drainage channels can be an expensive operation which must be regularly repeated. Good engineering maintenance of channels will considerably reduce the costs of mollus-ciciding
(Photo: R Witlin, World Bank)

The clearing of aquatic vegetation can also be effective, as most species of snail lay their eggs on leaves. It has been found in Tanzania that this simple measure can retard the reinfestation by snails of ponds and dams which have been treated with molluscicide.

Snail control measures requiring mobilization of the population face the difficulty that schistosomes are long-lived, so that it may be several years before benefits, in terms of a health improvement, are discernible to the inhabitants.

It is worth stressing that we still have much to learn abut the relationships between environmental changes and snail populations, and between snail populations and schistosomiasis transmission. There is much need for experiment and investigation. For instance, on existing irrigation schemes, a study of why one canal is full of snails and another has none may make it possible to control snails by altering conditions in the first canal to reproduce those in the second (Figure 17.8).

When snail control is attempted, it should be evaluated to assess its effectiveness. However, snail populations often vary with the seasons, and a seasonal drop in the snail population should not be mistaken for success in snail control. Rather, snail control measures should be timed to take effect at the most favourable season, usually the dry season when snail populations are at their lowest.

In many areas, schistosomiasis is a man-made disease. Not only irrigation schemes are liable to promote snail-breeding, but also the ditches, borrow pits, quarries, and pools carelessly created by road and railway construction, industrial activity, and so on. These can become important schistosomiasis transmission foci, especially when they are in or near towns and cities. The elimination of these bodies of water should be the direct responsibility of the agency which created them, and in most cases is not very difficult or expensive.

17.3 SCHISTOSOME REMOVAL FROM WATER AND WASTES

Introduction

Theory in this field is not well developed. The results of studies reported in the literature are not always comparable and some are contradictory. In general, studies have been made of egg and miracidia survival in sewage treatment and of cercariae survival in water treatment. Many of the high removal rates of schistosome eggs in sewage treatment are undoubtedly due to hatching in tests where miracidia were not enumerated in the effluent.

Figure 17.8 A snail survey in Egypt. The size of the snail held in the tweezers is typical for the vector species. A detailed knowledge of the ecology of the important vector snail species in a particular area is essential to designing any environmental control campaign (Photo: D Henrioud, WHO)

Cercariae in water treatment

Cercariae can be highly mobile and are not reliably removed by sedimentation with coagulation. Rapid sand filtration is ineffective and even slow sand filtration cannot guarantee the removal of all cercariae. Chlorination is effective against cercariae and, as with its bactericidal activity, it is more lethal at lower pH and warmer temperature. At 20°C, the concentrations of free-residual chlorine necessary to inactivate cercariae in thirty minutes have been reported as:

> 0.3 mg/l at pH 5.0;
>
> 0.6 mg/l at pH 7.5;
>
> 5.0 mg/l at pH 10.0.

Higher doses of chlorine need to be added, in order to obtain these residuals after meeting the immediate chlorine demand (Chapter 6).

Storage is highly effective against cercariae, which will be non-infective within forty-eight hours at warm temperatures (Figure 17.9).

Eggs and miracidia in sewage treatment

Intense schistosomiasis transmission may result from the discharge of effluents containing schistosome eggs to streams and lakes. Correct sewage treatment can prevent this. Different sewage treatment processes have different effects upon the survival of eggs and miracidia. A detailed account will be found in Feachem *et al.* (1983). In primary sedimentation some, but not all, eggs will settle out. Eggs are unlikely to hatch in sedimentation tanks but they can maintain their viability for at least twenty-four hours and hatch if transferred to a more favourable environment.

Eggs contained in sludge will die if anaerobically digested at above 25° C for twenty-four days. At cooler temperatures, longer periods will be required. On sludge drying beds, three weeks should be allowed for the elimination of hatchable eggs.

Trickling filters and activated-sludge plants provide a relatively favourable environment for eggs. Many will pass through and many will hatch. In an activated-sludge plant most eggs hatch in a few hours. This is a good thing because the miracidia die unless they find a snail within twenty-four hours.

Waste stabilization ponds with an adequate total retention time (at least twenty days) provide a total barrier to schistosomes. Anaerobic ponds strongly inhibit hatching of eggs while facultative and aerobic

Figure 17.9 Survival of *Schistosoma japonicum* cercariae in raw waters. Cercariae lose the ability to penetrate the skin and cause infection some time before they die. The line for loss of infectivity therefore lies below the line shown and roughly parallel to it
Source: From Jones and Brady (1947). Reproduced by permission of the US National Institutes of Health

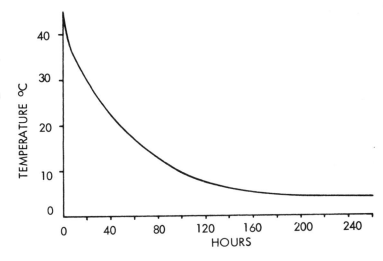

ponds do not. Miracidia die within six hours in anaerobic ponds and within ten hours in facultative ponds. In anaerobic ponds snails cannot reproduce and will die within a few weeks, while they can live and breed in facultative and maturation ponds. If a viable egg enters a facultative pond in a three-pond system, it may hatch and infect a snail; but any cercariae shed by the snail still have two maturation ponds to pass through and die long before they reach the outfall.

17.4 REFERENCES AND FURTHER READING

Araoz, J. de (1962). Study of water-flow velocities in irrigation canals in Iraq and their mathematical analysis, *Bulletin of the World Health Organization*, **27**, 99–123.

Bolton, P. (1988). Schistosomiasis control in irrigation schemes in Zimbabwe, *Journal of Tropical Medicine and Hygiene*, **91**, 107–114.

Feachem, R.G., Bradley, D.J., Garelick, H. and Mara, D.D. (1983). *Sanitation and Disease: Health Aspects of Excreta and Wastewater Management* (London: John Wiley).

Fenwick, A. (1972). The costs and cost-benefit analysis of an *S. mansoni* control programme on an irrigated sugar estate in Northern Tanzania, *Bulletin of the World Health Organization*, **47**, 573–578.

Frick, L.P. and Hillyer, G.V. (1985). The influence of pH and temperature on the cercaricidal activity of chlorine, *Military Medicine*, **131**, 372–378.

Jeffrey, H.C. and Leach, R.M. (1975). *Atlas of Medical Helminthology and Protozoology*, 2nd edition (Edinburgh: Churchill Livingstone).

Jobin, W.R. (1970). Control of *Biomphalaria glabrata* in a small reservoir by fluctuation of the water level, *American Journal of Tropical Medicine and Hygiene*, **19**, 1049–1054.

Jones, M.F. and Brady, F.J. (1947). Survival of *Schistosoma japonicum* cercariae at various temperatures in several types of water, in *Studies on Schistosomiasis*, US National Institutes of Health Bulletin No. 189 (Washington: US Government Printing Office).

Jordan, P. and Webbe G. (1982). *Schistosomiasis: Epidemiology, Treatment and Control* (London: Heinemann).

Kraatz, D.B. (1971). *Irrigation Canal Lining*, Irrigation and Drainage Paper No. 2 (Rome: Food and Agriculture Organization).

McCulloch, F. (1986). Snail control in relation to a strategy for reduction of morbidity due to schistosomiasis. *Tropical Medicine and Parasitology*, **37**, 181–184.

McJunkin, F.E. (1970). *Engineering Measures for the Control of Schistosomiasis* (Washington: USAID).

McJunkin, F.E. (1975). *Water, Engineers, Development and Disease in the Tropics* (Washington: USAID).

Muller, M. (1979). The engineer in the control of schistosomiasis, *Progress in Water Technology*, **11**, 167–172.

Pike, E.G. (1987). *Engineering against Schistosomiasis/Bilharzia* (London: Macmillan).

Unrau, G.O. (1979). Water supply and schistosomiasis in St. Lucia, *Progress in Water Technology*, **11**, 181–190.

WHO (1980). *Environmental Management for Vector Control*, Technical Report Series No. 649 (Geneva: World Health Organization).

WHO (1985). *Control of Schistosomiasis*, Technical Report Series No. 728 (Geneva: World Health Organization).

Wright, W.H. (1972). A consideration of the economic impact of schistosomiasis, *Bulletin of the World Health Organization*, **47**, 559–566.

Appendix A

Biological Classification Conventions

The scientific names for the species of living organisms follow a system based on a hierarchy of classificatory groups. Members of the plant and animal kingdom are divided into broad groups known as *Phyla*, such as the protozoa, the arthropods, or the molluscs. Each phylum is further divided into *Classes*, classes into *Orders*, and so on down through *Genera* to *Species*. For example, man, a filariasis-transmitting mosquito, and the typhoid bacterium are respectively classified as follows:

	Man	*Mosquito*	*Typhoid bacterium*
PHYLUM	Chordata	Arthropoda	Procaryotae
CLASS	Mammalia	Insecta	Schizomycetes
ORDER	Primates	Diptera	Pseudomonadales
FAMILY	Hominidae	Culicidae	Enterobacteriaceae
GENUS	*Homo*	*Culex*	*Salmonella*
SPECIES	*Homo sapiens*	*Culex quinquefasciatus*	*Salmonella typhi*

Sometimes there are intermediate categories between the levels of this hierarchy, such as the culicine and anopheline subfamilies of the Culicidae family (see Appendix B). Each species is technically referred to by a name indicating both its genus and species; for instance *Homo sapiens*. The first (generic) name is written with an initial capital and the second (specific) name in lower-case, and both in italics.

To indicate a subspecies or variety of an organism, a third name is added. For instance, the variety of louse found on the scalp is indistinguishable in appearance from that found on the body, and to differentiate them they are known as *Pediculus humanus capitis* and *Pediculus humanus corporis*, respectively.

Generic names are often abbreviated, as in *S. typhi* for *Salmonella typhi*. Confusion may arise from these abbreviations. *E. coli* normally refers to the intestinal bacterium *Escherichia coli*, but may refer to the intestinal protozoon *Entamoeba coli*. Where such confusion is possible it is preferable to abbreviate the latter to *Ent. coli*. Several species of one genus, such as *Anopheles*, may be referred to by an abbreviation of the form: *Anopheles* spp. A single, but unidentified species is referred to as *Anopheles* sp.

The same conventions apply to the organisms which infect animals as to the animals themselves. If infection with an organism causes a disease, the organism is a pathogen, and the disease may be named after it. The name of the disease sometimes suggests the type of pathogen involved. For instance, the names of diseases caused by parasites such as protozoa and worms (helminths) frequently end in -iasis; thus, helminths of the genus *Schistosoma* cause schistosomiasis. Similarly, infections caused by bacteria and some other organisms may have the ending -osis. Thus, bacteria of the genus *Shigella* cause shigellosis. The plural of -iasis is -iases and the plural of -osis is -oses.

Appendix B

Glossary

NOTE: Words in **bold** type are also explained in this glossary.

activated sludge: A common method of biological **sewage** treatment. Settled sewage is led into an aeration tank where oxygen is supplied either by mechanical agitation or by diffused aeration. The **bacteria** which grow in this **medium**, together with other solids, are removed in a secondary **sedimentation** tank and recycled to the aeration tank inlet. This creates a high concentration of biologically active **flocs** in the aeration tank.

aerobic: Living or taking place in the presence of air or oxygen.

aestivate: To pass the summer (or dry season) in a dormant state.

alluvium: Loose silt or clay deposited beside a river.

algae: Primitive plants, one- or many-celled, usually **aquatic** and capable of photosynthesis.

anaerobic: Living or taking place without air or oxygen.

anopheline: Belonging to the anophelini subfamily of mosquitoes, which includes the genus *Anopheles*.

aqua-privy: A toilet with a sealed settling chamber located directly underneath, receiving only **excreta** and small volumes of flushing water through a chute. **Retention times** may be up to sixty days and **effluent** goes to a **soakaway** or to small-diameter **sewers**.

aquatic: Living in water.

arboviruses: **Viruses** carried by **arthropods**.

arthropod: A jointed-limbed invertebrate animal belonging to the **phylum** Arthropoda, which includes the insects, **crustaceans**, and spiders.

ASCE: American Society of Civil Engineers.

bacillus (plural: *bacilli*): A rod-shaped **bacterium**.

bacterium (plural: *bacteria*): **Micro-organism** of simple structure and small size (1–10 μm), intermediate between the plant and animal kingdoms and commonly rod-shaped (**bacilli**, e.g. *Escherichia*) or round (cocci, e.g. *Streptococci*).

biodegradable: Capable of being broken down by biological processes.

biogas: Gas consisting mainly of methane produced by **anaerobic digestion** of organic waste.

BOD: Biochemical oxygen demand, defined in Section 3.2.

borrow pit: A pit formed by the removal of earth for construction or other purposes.

carrier: An infected person (or animal) that harbours a specific **pathogen** in the absence of discernible clinical disease and serves as a potential source of infection for man.

cartage: Systems of **nightsoil** removal involving vehicular or manual removal. For instance, bucket latrine emptying into carried containers, carts, or trucks and vault emptying by suction pumps into tankers.

cercaria (plural: *cercariae*): The **larval** stage of a **trematode** worm which emerges from the snail **host**. Often refers to the final larval stage of schistosome species, which leaves an **aquatic** snail and infects man through the skin.

cestodes: Tapeworms of the class Cestoda. Many cestodes have an adult stage in the intestine of one **host** (e.g. man) and an encysted stage in the flesh of another (e.g. cow); e.g. *Taenia saginata*

chemotherapy: Medical treatment by chemical means.

COD: Chemical oxygen demand, the mass of oxygen consumed when the **organic** matter present is oxidized by strong oxidizing agents in acid solution.

coliforms: A group of **bacteria**. Some of them, faecal coliforms, are normally found in human and animal faeces. They are Gram-negative, **aerobic** and facultatively **anaerobic**, non-spore-forming rods which grow in the presence of bile salts and ferment lactose producing acid and gas.

community: A collection of different species living in a particular environment.

compost: The humus-like product of the **aerobic** or **anaerobic** composting of either **nightsoil** or **sludge** mixed with **organic** material rich in carbon (such as refuse or sawdust).

composting toilet: A toilet into which **excreta** and carbon-rich material are added (vegetable wastes, straw, grass, sawdust, ash), and special conditions maintained to produce an inoffensive **compost**.

copepods: An order of the class **Crustacea** of the **phylum Arthropoda**; e.g. *Cyclops*.

crustacea: A large class of **arthropods** including crabs, lobsters, shrimps, and water fleas.

culicine: Belonging to the culicini subfamily of mosquitoes, which includes the **genera** *Culex, Aedes*, and *Mansonia*.

cutaneous: Related to the skin.

desludging: Removing accumulated **sludge** from septic tanks, **aqua-privies**, etc.

digestion: The breaking down of **organic** waste by **bacteria**.

drawdown: Lowering the water level behind a dam by releasing water.

ecology: The study of the relationships between **communities** of organisms and their environment.

effluent: Outflowing liquid.

endemic: Describes a disease or **pathogen** constantly present within a given geographic area or community.

epidemic: The occurrence in a community or region of cases of an illness (or an outbreak) clearly in excess of normal expectancy.

epidemiology: The study of the geography, frequency, environmental and behavioural causes, and transmission of disease.

eutrophication: The enrichment of natural waters, especially by compounds of nitrogen and phosphorous, resulting in increased productivity of some species of plants.

excreta: In this book excreta refers to faeces and urine.

facultative pond: A pond which is **aerobic** near the surface, but **anaerobic** lower down.

fall: Slope along a pipe or channel.

fatality rate: Usually expressed as a percentage of the number of persons diagnosed as having a specified disease who die as a result of that illness.

faecal–oral: Transmitted by any route enabling faecal material to reach the mouth.

fissured: Containing fissures or cracks.

flagellate: A minute, single-celled animal (**protozoon**) able to swim with one or more whip-like structures (flagella); e.g. *Giardia*.

floc: Agglomeration of small particles suspended in water.

fluke: A **parasitic** flatworm of the **phylum** Platyhelminthes, class Trematoda, usually having a snail intermediate **host**, e.g. *Schistosoma*.

focus: A point at which intense disease transmission occurs.

genus (plural: *genera*): See Appendix A.

groundwater: Water located beneath the ground surface.

habitat: A place or environment in which an organism naturally prospers and breeds.

haemorrhagic: Causing haemorrhage or bleeding.

helminth: A worm; the helminths discussed in this book are **parasitic** worms; e.g. *Ascaris, Schistosoma*, and *Taenia*.

host: See '**parasite**'.

immunity: A capacity to resist **infection** by a particular **pathogen**, acquired by previous infection or vaccination.

impounding: Filling a reservoir by damming a river.

incidence: The number of cases of a specified disease diagnosed or reported during a defined period of time, divided by the number of persons in the population in which they occurred.

infection: An infectious disease (an infection) is one which is caused by a **pathogenic** organism, and can therefore be passed from one person to another. A person may be infected with a pathogen without suffering the symptoms of the disease.

infective dose: The number of **pathogens** which must simultaneously enter the body, on average, to cause infection.

ingest: To take into the body by swallowing.

intestinal tract: The part of the alimentary canal or digestive tube beyond the stomach.

invert: The lowest point on the internal surface of a channel or pipe.

larva (plural: *larvae*): A stage in the development of some organisms, including **helminths** and insects, differing from the embryo in that it can secure its own nourishment.

leachate: Water draining from or through soil or a **refuse** pile or some other material.

leaching pit: A kind of **soakaway**.

lymphatic system: A system of small ducts which return excess fluid from the body's tissues to the bloodstream.

maturation ponds: The final ponds in a series of waste stabilization ponds. They are entirely **aerobic**.

medium: A substance in which **bacteria** can multiply. Laboratories use special media in which only selected species thrive.

metabolic: Related to the processes of chemical change which take place in living organisms, such as digestion and respiration.

mg/l: Milligrammes per litre; 1 mg/l is roughly equivalent to 1 part per million.

micro-organism: A microscopically small organism.

miracidia (singular: *miracidium*): The embryos of **trematodes**. Often refers to schistosome embryos which invade the bodies of snails.

molluscicide: A chemical which kills molluscs such as snails.

morbidity: The showing of symptoms of disease.

nematodes: Roundworms and other similar worms of the class Nematoda; e.g. *Ascaris*.

nightsoil: Human **excreta** transported without flushing water.

oral: Of the mouth.

organic: Derived from living material or, in the case of chemicals, containing carbon.

organochlorides: **Organic** chemicals containing chlorides.

ovum (plural: *ova*): An egg.

parasite: An organism that lives on or in another living organism, termed the **host**, and draws nourishment from it.

pathogen: A pathogen or pathogenic organism is an organism which causes disease. Most pathogens are microscopic in size.

peak factor: The ratio of maximum likely water use in a peak period to the average rate of water use.

percolation: The soaking of liquids through the soil.

pH: A measure of acidity or alkalinity, which can take values between 0 (extremely acid) and 14 (extremely alkaline).

phenols: **Toxic organic** chemicals formed by the breakdown of various oil-based chemicals such as petrol and bitumen.

phylum: See Appendix A.

prevalence: The number of persons sick or portraying a certain condition at a particular time (regardless of when that illness or condition began) divided by the number of persons in the population in which they occurred.

priming: Filling a suction pump or a siphon to enable it to operate.

protoplasm: The characteristic material of all living tissue.

protozoon (plural: *protozoa*): The smallest and simplest creatures that can be called animals. Each has only one cell and is between 0.002 and 0.5 mm in size; e.g. *Entamoeba histolytica*.

puddled clay: Clay mixed with water, and sand if necessary, placed in 150 mm layers and kneaded into place to form a watertight seal.

pupa (plural: *pupae*): The non-feeding stage in an insect's development between **larva** and adult.

pupate: To develop into a **pupa**.

refuse: Rubbish or garbage.

reservoir: Any human beings, animals, **arthropods**, plants, soil, or inanimate matter in which a **pathogen** normally lives and multiplies, and on which it depends primarily for survival. For instance, man is the only reservoir of typhoid bacteria.

retention time: The period of time which water or wastes take to pass through a septic tank, stabilization pond, or other process.

rickettsia: Simple single-celled organisms, smaller than **bacteria** but larger than **viruses** (0.3–2.0 μm). They are **parasites** and grow in the living tissue of an appropriate **host**; e.g. *Rickettsia prowazeki*, the cause of louse-borne typhus.

rip-rap: An arrangement of large stones or pieces of concrete to prevent wave erosion on the banks of a lagoon, reservoir, or dam.

scum: Solid material, with fats, oils, grease, and soaps, floating on the surface of a septic tank, pond, etc. Scum often forms large floating masses, called scum mats.

sedimentation: The process by which suspended solid particles in water or **sewage** are allowed to settle out to the bottom of a tank or pond.

sewage: Human **excreta** and waste water, flushed along a **sewer** pipe.

sewer: A pipe containing waste water or **sewage**.

sewerage: A system of **sewer** pipes.

sludge: A mixture of solids and water deposited on the bottom of septic tanks, ponds, etc.

soakaway: An arrangement to promote seepage of **effluent** into the ground.

species sanitation: The control of disease **vectors** by methods directed at the behaviour of particular species.

spillway: The arrangement for release of excess water from a dam when the reservoir is full.

spirochaetes: An order of **bacteria** with slender, spiral bodies; e.g. *Leptospira*.

spoil: Soil which has been removed by digging a pit or trench.

sullage: Domestic dirty water not containing **excreta**, also called gray water.

taxonomy: The science of classification (see Appendix A).

toxic: Poisonous.

transpiration: Evaporation of moisture from the leaves of a plant; the moisture came from the soil, diffused into the roots, and moved up the stem to the leaves.

trematodes: Flat worms of the class Trematoda, including the **parasitic** worms called **flukes**. Trematodes of medical importance have intermediate stages in snails; e.g. *Schistosoma*.

vector: Animal - often an insect - transmitting an **infection** from person to person or from infected animals.

vertebrate: Belonging to the **phylum** Chordata; this includes fish, reptiles, birds and mammals.

virus: An exceedingly small **parasitic** organism (0.01–0.3 μm). Viruses can only reproduce inside the animal or plant cells of a suitable **host**, but some of them can survive for long periods elsewhere; e.g. hepatitis A virus.

visceral: Related to the viscera, the organs in the abdomen such as the heart, liver, and intestines.

water table: The level in the ground at which water is found when a hole is dug or drilled.

worm load: The number of **parasitic** worms (**helminths**) with which a person is **infected**.

zoonosis: An **infection** or an infectious disease transmissible under natural conditions from vertebrate animals to man. May be enzootic or epizootic (which have meanings analogous to **endemic** and **epidemic** respectively).

Appendix C

Checklist of Water-related and Excreta-related Diseases

Disease	Common name	Pathogen	Transmission	Distribution	Water-related category (Table 1.2)	Excreta-related category (Table 1.3)
A. Bacterial Diseases						
Bacterial enteritis	Diarrhoea, gastroenteritis	*Campylobacter jejuni*, *Escherichia coli*, *Salmonella* spp, *Yersina enterocolitica*	Faecal–oral, man → man or animal → man	Worldwide, particularly serious and common among children	1	II
Shigellosis	Bacillary dysentry	*Shigella* spp.	Faecal-oral, man → man	Worldwide	1	II
Cholera	Cholera	*Vibrio cholerae*	Faecal–oral, man → man	Worldwide	1	II
Paratyphoid	Paratyphoid	*Salmonella paratyphi*	Faecal–oral, man → man	Worldwide	1	II
Typhoid or enteric fever	Typhoid	*Salmonella typhi*	Faecal–oral man → man	Worldwide	1	II
B. Spirochaetal Diseases						
Leptospirosis	Weil's disease	*Leptospira* spp.	Excreted by animals (notably rodents) in urine and infects humans through skin, mouth, or eyes; animal → man	Worldwide	1	—
Louse-borne relapsing fever		*Borrelia recurrentis*	Louse-borne, man → louse → man	Worldwide but mainly in poor mountainous parts of Africa. Asia, and Latin America	2(b)	—

Disease	Common name	Pathogen	Transmission	Distribution	Water-related category (Table 1.2)	Excreta-related category (Table 1.3)
C. Viral Diseases						
C1 Excreted viruses						
Hepatitis A	Infectious hepatitis or jaundice	Hepatitis A virus	Faecal–oral, man → man	Worldwide	1	I
Poliomyelitis	Polio	Poliovirus	Faecal–oral, man → man	Worldwide	1	I
Viral diarrhoea	Diarrhoea	Rotavirus, Norwalk agent, other viruses	Faecal–oral, man → man	Worldwide	1	I
C2 Mosquito-borne viruses						
Dengue	Breakbone fever	Dengue virus	Transmitted by mosquito *Aedes aegypti* and other *Aedes* spp., man → mosquito → man	Dengue fever is now almost worldwide. A new serious form (dengue haemorrhagic fever) occurs mainly in the cities of S.E. Asia	4(b)	—
Yellow fever		Yellow fever virus	Transmitted by mosquito *Aedes aegypti*, and other *Aedes* and *Haemagogus* spp., man or monkey → mosquito → man	Not reported from Asia or Australasia	4(b)	—
Other arboviral diseases		A large number of viruses causing various encephalitic and haemorrhagic infections	Mainly infections of animals and all transmitted by arthropods. Man infected accidentally by bites from mosquitoes (mainly), ticks, sandflies, and midges	Worldwide	4(b)	—

Disease	Common name	Pathogen	Transmission	Distribution	Water-related category (Table 1.2)	Excreta-related category (Table 1.3)
D. Rickettsial Disease						
Louse-borne typhus	Epidemic or classical typhus	*Rickettsia prowazeki*	Louse-borne, man → louse → man	Worldwide but mainly in poor mountainous areas of Europe, Asia, Africa, and Latin America	2(b)	—
E. Protozoal Diseases						
E1 Excreted protozoa						
Amoebiasis	Amoebic dysentery	*Entamoeba histolytica*	Faecal–oral, man → man	Worldwide	1	I
Balantidiasis	Diarrhoea	*Balantidium coli*	Faecal–oral, man or pig → man	Worldwide	1	I
Crytosporidiosis	Diarrhoea	*Cryptosporidium* spp.	Faecal–oral, man or animal → man	Worldwide	1	I
Giardiasis	Diarrhoea	*Giardia lamblia*	Faecal–oral, man → man	Worldwide	1	I
E2 Vector-borne protozoa						
Malaria	Malaria	*Plasmodium* spp.	Transmitted by *Anopheles* mosquitoes, man → mosquito → man	Throughout most of the warmer parts of the world, although eradicated in some areas (Figure 15.2)	4(b)	—
Trypanosomiasis (African)	Gambian sleeping sickness	*Trypanosoma gambiense*	Transmitted by riverine tsetse fly (*Glossina* spp.), man → fly → man	Mainly West and Central Africa	4(a)	—
	Rhodesian sleeping sickness	*Trypanosoma rhodesiense*	Transmitted by game tsetse fly (*Glossina* spp.), wild game or cattle → fly → man	Mainly East Africa	Not related to water or excreta (included for completeness)	
Trypanosomiasis (American)	Chagas' disease	*Trypanosoma cruzi*	Transmitted by bugs (Reduviidae), man or animal → bug → man	Latin America	Related to housing	

Disease	Common name	Pathogen	Transmission	Distribution	Water-related category (Table 1.2)	Excreta-related category (Table 1.3)

F. Helminthic Diseases

F1 Excreted helminths

Disease	Common name	Pathogen	Transmission	Distribution	Water-related category (Table 1.2)	Excreta-related category (Table 1.3)
Ascariasis	Roundworm	*Ascaris lumbricoides*	man → soil → man (see Appendix D)	Worldwide	—	III
Clonorchiasis	Chinese liver fluke	*Clonorchis sinensis*	animal or man → aquatic snail → fish → man (see Appendix D)	S.E. Asia	3(b)	V
Diphyllobothriasis	Fish tapeworm	*Diphyllobothrium latum*	man or animal → copepod → fish → man	Worldwide	3(b)	V
Enterobiasis	Pinworm	*Enterobius vermicularis*	man → man	Worldwide	1	I
Fasciolopsiasis	Giant intestinal fluke	*Fasciolopsis buski*	man or pig → aquatic snail → aquatic plant → man	S.E. Asia, mainly China	3(b)	V
Hymenolepiasis	Dwarf tapeworm	*Hymenolepis nana*	man or rodent → man	Worldwide	1	I
Hookworm	Hookworm	*Ancylostoma duodenale, Necator americanus*	man → soil → man(see Appendix D)	Mainly in warm, wet climates	—	III
Opisthorchiasis	Cat liver fluke	*Opisthorchis felineus, O.viverrini*	Cat or man → aquatic snail → fish → man	Thailand, USSR	3(b)	V
Paragonimiasis	Lung fluke	*Paragonimus westermani*	Pig, man, dog, cat, or animal → aquatic snail → crab or crayfish → man	East Asia plus scattered foci in Africa and S. America	3(b)	V

Disease	Common name	Pathogen	Transmission	Distribution	Water-related category (Table 1.2)	Excreta-related category (Table 1.3)
Schistosomiasis	Bilharziasis	*Schistosoma haematobium*	man → aquatic snail → man	Africa, Middle East, and India (Figure 17.1)	3(a)	V
		S. mansoni	man → aquatic snail → man	Africa, Middle East, and Latin America (Figure 17.2)	3(a)	V
		S. japonicum	animals and man → snail → man	S.E. Asia (Figure 17.1)	3(a)	V
Strongyloidiasis	Threadworm	*Strongyloides stercoralis*	man → soil → man	Mainly in warm, wet climates	—	III
Taeniasis	Beef tapeworm	*Taenia saginata*	man → cow → man (see Appendix D)	Worldwide	—	IV
	Pork tapeworm	*Taenia solium*	man → pig → man	Worldwide	—	IV
Trichuriasis	Whipworm	*Trichuris trichiura*	man → soil → man	Worldwide	—	III

F2 Guinea worm

Disease	Common name	Pathogen	Transmission	Distribution	Water-related category (Table 1.2)	Excreta-related category (Table 1.3)
Dracunculiasis (dracontiasis)	Guinea worm	*Dracunculus medinensis*	man → *Cyclops* → man (see Appendix D).	Africa, India (Figure 1.3)	3(b)	—

Disease	Common name	Pathogen	Transmission	Distribution	Water-related category (Table 1.2)	Excreta-related category (Table 1.3)
F3 Mosquito-borne helminths						
Filariasis (Bancroftian)	May cause elephantiasis	*Wuchereria bancrofti*	Transmitted by mosquitoes, mainly *Culex pipiens, Anopheles* spp., and *Aedes* spp., man → mosquito → man	Worldwide (Figure 15.4)	4(b)	VI
Filariasis (Malayan)	May cause elephantiasis	*Brugia malayi*	Transmitted by mosquitoes, mainly *Mansonia* spp. and also *Anopheles* and *Aedes* spp., man → mosquito → man	India and S.E. Asia	4(b)	—
Loiasis		*Loa loa*	Transmitted by mangrove fly (*Chrysops* spp.) man → fly → man	West and Central Africa	4(b)	—
Onchocerciasis	River blindness	*Onchocerca volvulus*	Transmitted by blackflies (*Simulium* spp.) man → fly → man	Latin America, Africa, and Yemen (Figure 15.7)	4(b)	—

Appendix D

The Life Cycles of Certain Helminths Infecting Man

The life cycle of the schistosomes is illustrated in Figure 17.3

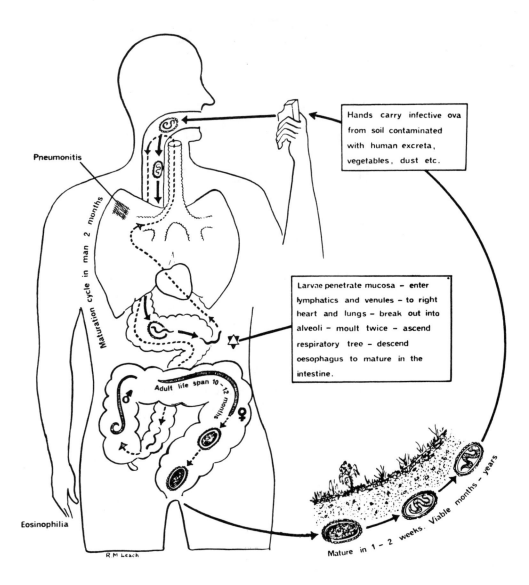

Figure D1   The life cycle of *Ascaris lumbricoides* (the round worm). *Trichuris trichiura* (whipworm) has a similar life cycle
(*Source*: From Jeffrey and Leach (1975). Reproduced by permission of Churchill Livingstone)

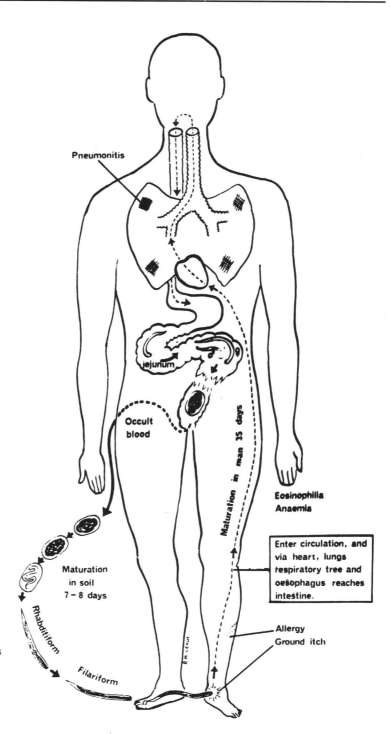

Figure D2 The life cycle of the hookworms (*Source*: From Jeffrey and Leach (1975). Reproduced by permission of Churchill Livingstone)

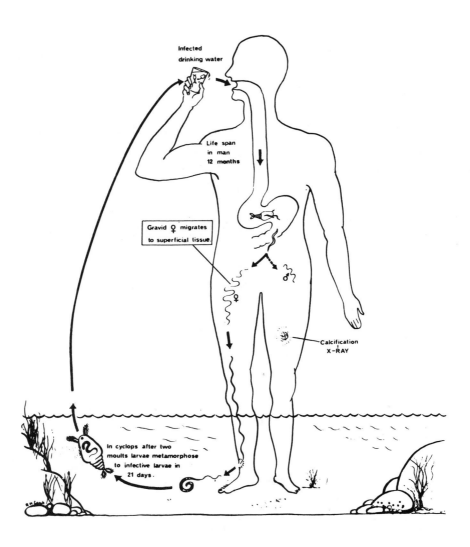

Figure D3   The life cycle of *Dracunculus medinensis* (the Guinea worm)
(*Source*: From Jeffrey and Leach (1975). Reproduced by permission of. Churchill Livingstone)

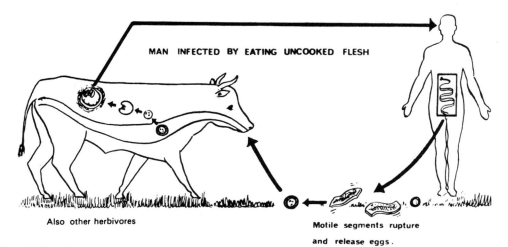

Figure D4 The life cycle of *Taenia saginata* (the beef tapeworm)
(*Source*: From Jeffrey and Leach (1975). Reproduced by permission of Churchill Livingstone)

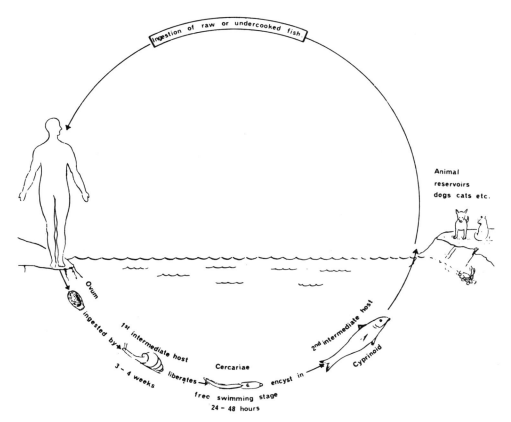

Figure D5 The life cycle of *Clonorchis sinensis* (the oriental liver fluke)
(*Source*: From Jeffrey and Leach (1975). Reproduced by permission of Churchill Livingstone)

Appendix E

Units

Quantities are given in this book in metric units. Approximate equivalents in Imperial and US units are as follows.

	Metric	*Imperial equivalent*
Length	1 mm = 1000 μm	0.04 inches
	25 mm	1 inch
	0.1 m = 10 cm = 100 mm	4 inches
	0.3 m = 30 cm = 300 mm	12 inches = 1 foot
	1 m = 100 cm = 1000 mm	3 feet = 1 yard
	1 km	0.6 miles
Area	1 m^2 (1 square metre)	1.2 square yards
	1 ha = 10 000 m^2	2.5 acres
Volume	1 litre = 1000 ml	1.8 Imperial pints, or 2.1 US pints
	10 l	2.2 Imperial gallons, or 2.6 US gallons
	1 m^3 = 1000 l	220 Imperial gallons, or 264 US gallons
Weight	30 g = 30 000 mg	1 ounce
	1 kg = 1000 g	2.2 pounds (lbs)
Density	1 kg/l	62 lbs/cubic foot
Concentration	1 mg/l	1 part per million
	1 μg/l	1 part per billion

Index